DEPARTMENT OF ENERGY

The Public Inquiry into the Piper Alpha Disaster

The Hon Lord Cullen

VOLUME TWO

Presented to Parliament
by the Secretary of State for Energy
by Command of Her Majesty
November 1990

LONDON: HMSO

Cm 1310

£38.00 net
(Two volumes not sold separately)

SECTION FOUR: THE FUTURE

LIST OF PLATES

APPENDICES

1. Piper Alpha platform: east and west elevations
2. Piper Alpha platform: south and north elevations
3. Layout of the production deck and Modules A and B
4. Layout of Modules C and D, Module D mezzanine level, and SubmoduleD
5. The Piper Alpha platform, showing the pipe deck and other features related to the escape from the platform
6. The 68 ft level, or deck support frame
7. Layout of the accommodation modules
8. Simplified flow diagram for phase 1 operation on 6 July
9. The condensate injection pump system
10. The gas detection system in C Module

Printed in the United Kingdom for Her Majesty's Stationery Office
Dd 0503434/11/90, C75

SECTION FOUR: THE FUTURE

Chapter 16

Offshore and Onshore Safety Regimes

Introduction

16.1   In this chapter I will give by way of background to the following chapters a brief outline of the existing UK offshore safety regime and, since comparison with them has been of assistance to me, the UK onshore safety regime and the Norwegian offshore safety regime. A complete account is not practicable. I will concentrate on aspects which were of most relevance to the issues discussed in Part 2 of the Inquiry. I heard the evidence of Mr J R Petrie, Director of Safety, PED, and Mr R J Priddle, Deputy Secretary at the Department of Energy since September 1989; Mr J Rimington, Director General of the HSE and Mr D J Hodgkins, Director of the Safety and General Policy Division, HSE; and Mr M Ognedal, Director of the Safety and Working Environment Division, NPD since 1980; in addition to the evidence in Part 1 of Mr F H Atkinson, Manager of the Offshore Division, Lloyd's Register of Shipping. This chapter ends with some remarks on future trends in offshore operations in the North Sea, in the light of evidence given by Dr B G S Taylor, Director of Technical Affairs, UKOOA.

The UK offshore safety regime

16.2   Exploration and production licences are subject to model clauses prescribed by regulations under the Petroleum (Production) Act 1934 (PPA) as applied to the areas of the continental shelf which are subject to UK jurisdiction. The clauses which are concerned with safety require licensees to (i) obtain the consent of the Minister to the drilling or abandonment of a well; (ii) execute all operations in accordance with good oilfield practice; and (iii) comply with instructions given by the Minister for the health, safety and welfare of persons employed in or about the area to which the licence relates. However, in the light of the MWA, which was specifically designed to cater for offshore safety, the main significance of the clauses is in regard to the minimising of the risk of a well "blowout". The regulations also require the identification and approval of the "operator" which for practical purposes is to discharge the responsibilities of the licensees.

The Mineral Workings (Offshore Installations) Act 1971 (MWA)

16.3   The MWA was enacted as a consequence of the investigation of the collapse of the exploration rig "Sea Gem" in 1965 and the recognition that arrangements based on the PPA were not appropriate for the purposes of securing offshore safety. The MWA, as amended, *inter alia*:

(i)   required the registration of offshore installations (Sec 2).

(ii)   empowered the Secretary of State to make regulations "requiring offshore installations or parts of offshore installations to be certified by such persons and in such manner as may be provided by the regulations to be, in respect of such matters affecting safety as may be so provided, fit for the purpose or purposes specified by the regulations ..." (Sec 3(1)).

(iii)   empowered him to make regulations "for the safety, health and welfare of persons on offshore installations ... and generally, and whether or not by way of supplementing the preceding sections of this Act, for the safety of such installations and the prevention of accidents on or near them"; and to appoint as inspectors to discharge the functions conferred by the regulations, and generally to assist him in the execution of the Act, "such number of persons appearing to him to be qualified for the purpose as he may from time to time consider necessary or expedient ...". (Sec 6).

(iv) required the appointment of an OIM who was given a general responsibility for safety, health and welfare and for maintaining discipline and order (Secs 4 and 5).

16.4 Under the MWA an "offshore installation" includes any floating structure or device maintained on station by whatever means; and any pipeline works or apparatus deemed to form part of it, such as those covered by the Included Apparatus or Works Order; but did not otherwise include any part of a pipeline (Sec 1(5)). The MWA also made provision for a power on the part of the Secretary of State to give directions for the exemption of installations from the operation of regulations made under it (Sec 7).

16.5 Prior to the disaster a number of sets of regulations on specific subjects had been made. The most relevant for present purposes were the Inspectors and Casualties Regulations in 1973; the Construction and Survey Regulations in 1974; the Operational Safety, Health and Welfare Regulations and the Emergency Procedures Regulations in 1976; the Life-saving Appliances Regulations in 1977; the Fire Fighting Equipment Regulations in 1978; and the Well Control Regulations in 1980.

The Petroleum and Submarine Pipe-lines Act 1975 (PSPA)

16.6 A separate regulatory system was set up for offshore pipelines by the PSPA in 1975, under which the construction and use of such pipelines required authorisation by the Secretary of State. This involved considerations of planning and safety. He was also empowered to make regulations for the safe construction and operation of pipelines and the safety, health and welfare of pipeline workers; and the appointment of inspectors to enforce the regulations. Prior to the disaster regulations had been made under the PSPA on a number of subjects including diving operations and inspectors.

The Health and Safety at Work etc Act 1974 (HSWA)

16.7 In the meantime the HSWA was passed in 1974. This Act arose out of the report of the Robens Committee in 1972 (Cmnd 5034). That committee identified a number of major defects in the then existing statutory system for the advancement of safety and health. According to them the first and perhaps most fundamental was that there was too much law. This had had the unfortunate effect of conditioning people to think of safety and health at work as in the first and most important instance a matter of detailed rules imposed by external agencies.

"The matter goes deeper. We suggested at the outset that apathy is the greatest single contributing factor to accidents at work. This attitude will not be cured so long as people are encouraged to think that safety and health at work can be ensured by an ever-expanding body of legal regulations enforced by an ever-increasing army of inspectors. The primary responsibility for doing something about the present levels of occupational accidents and disease lies with those who create the risks and those who work with them. The point is quite crucial. Our present system encourages rather too much reliance on state regulation, and rather too little on personal responsibility and voluntary, self-generating effort. This imbalance must be redressed. A start should be made by reducing the sheer weight of the legislation. There is a role in this field for regulatory law and a role for Government action. But these roles should be predominantly concerned not with detailed prescriptions for innumerable day-to-day circumstances but with influencing attitudes and with creating a framework for better safety and health organisation and action by industry itself." (para 28).

The second main defect was that too much of the existing law was intrinsically unsatisfactory. The committee referred to problems created by unintelligibility and obsolescence. Further, the great bulk of the existing provisions were concerned with physical circumstances.

"But it has long been widely accepted that equally important factors in safety and health at work are the attitudes, capacities and performance of people and the efficiency of the organisational systems within which they work. This is not yet adequately reflected in the legislation. As a result, much of the legislation appears irrelevant to the real, underlying problems." (para 30).

A third major problem area identified by the committee was the fragmentation of administrative jurisdictions and the absence of a clear and comprehensive system of official provision for safety and health at work. As regards the objectives of future policy the committee observed:

"The most fundamental conclusion to which our investigations have led us is this. There are severe practical limits on the extent to which progressively better standards of safety and health at work can be brought about through negative regulation by external agencies. We need a more effectively self-regulating system. This calls for the acceptance and exercise of appropriate responsibilities at all levels within industry and commerce. It calls for better systems of safety organisation, for more management initiatives, and for more involvement of work people themselves. The objectives of future policy must therefore include not only increasing the effectiveness of the state's contribution to safety and health at work but also, and more importantly, creating conditions for more effective self-regulation." (para 41).

16.8 The HSWA made provision with a view to the progressive replacement of specific Acts and instruments relating to health and safety by a system of regulations and approved codes of practice. It imposed on an employer a duty "to ensure, so far as is reasonably practicable, the health, safety and welfare at work of all his employees" and "to conduct his undertaking in such a way as to ensure, so far as is reasonably practicable, that persons not in his employment who may be affected thereby are not thereby exposed to risks to their health or safety" (Secs 2 and 3). It established the Health and Safety Commission (HSC) as the body responsible for effecting the general purposes of the Act; and it established the Health and Safety Executive (HSE) as the body generally responsible for the enforcement of health and safety legislation and for exercising on behalf of the HSC such of its functions as the HSC directed it to exercise. For the purposes of enforcement provision was made for the appointment of inspectors, whose powers included the service of improvement and prohibition notices. The province of the work of the HSE excluded certain areas of industrial and technological hazard such as consumer and food safety and transport other than that of hazardous goods.

16.9 So far as health and safety offshore were concerned there were two differences. First, the HSWA did not apply outside Great Britain until an Order in Council so provided. Second, the HSWA did not treat the MWA or regulations made under it as part of the legislation which was subject to progressive replacement, despite the fact that the Robens Committee thought that they should be brought within the unified system "perhaps as a second stage after the main arrangements have been tackled - unless very sound reasons can be adduced for leaving them outside". (Para 109). It appears that this exclusion may have been influenced by a comparison between ships and installations, and in particular mobile installations, in respect of the hazards to which they were subject.

16.10 On 30 July 1976 the Prime Minister made a statement that the Government had decided that the HSWA should be extended to cover workers engaged in the offshore oil and gas industry, including divers, so that one agency, the HSC, would be responsible for ensuring that common standards of occupational safety were applied both on and offshore. However, in view of the knowledge and experience developed by the DEn (which had come into separate existence in 1974) on the technical aspects of structural safety and "blowout" risks, the Secretary of State for Energy would retain his existing responsibilities for safeguarding offshore workers against such dangers. The responsibility for inspecting offshore installations would remain with the PED, which would act as the agent of the HSC as regards occupational safety.

16.11   The HSWA (except Part III) was extended to the UK territorial waters and the UKCS by an Order in Council in 1977 (in 1989 superseded by a similar Order). However the regulations made under the HSWA were not to apply there except in so far as the regulations expressly so provided (Sec 15(9) of the HSWA). Prior to the disaster 7 sets of regulations dealing with particular types of hazard were given this extended effect. The only set of regulations which has so far been made under the HSWA which applies only to offshore operations is the Offshore Installations and Pipeline Works (First Aid) Regulations 1989.

16.12   Following the Prime Minister's statement the HSC in pursuance of its powers under Sec 13(1)(a) of the HSWA entered into an agency agreement with the Secretary of State for Energy dated 1 November 1978.

The report of the Burgoyne Committee

16.13   In view of the increasing level of offshore activity the Secretary of State for Energy in 1978 appointed a committee under Dr J H Burgoyne "To consider so far as they are concerned with safety, the nature, coverage and effectiveness of the Department of Energy's regulations governing the exploration, development and production of oil and gas offshore and their administration and enforcement. To consider and assess the role of the certifying authorities. To present its report, conclusions and any recommendations as soon as possible." In practice the Committee found it necessary to consider the work of other bodies under other legislation, and in particular the HSWA. In their report, which was submitted in 1980 (Cmnd 7866), the Committee made the following recommendations under the heading of "Administration and Enforcement":

"6.5 The Government shall discharge its responsibility for offshore safety via a single Government agency whose task is to set standards and ensure their achievement (4.10).

6.6 We consider that the Department of Energy is capable of discharging this responsibility effectively, provided it is suitably strengthened and seeks advice from other bodies on matters of common concern. The strengthening is to provide the ability to monitor and where necessary set safety standards in relation to the selection, training and qualification of offshore personnel (5.130), and to acquire additional expertise in matters of occupational safety generally (4.24). The principal sources of advice to which we refer are the Department of Trade on marine safety, the Civil Aviation Authority on aviation safety and the HSE on occupational safety (4.11).

6.7 We recommend that the Department of Energy should continue its policy to employ an inspectorate consisting of well-qualified and industrially-experienced individuals, capable of a broad but authoritative approach to their monitoring and enforcement functions (4.39). We further recommend that inspectors should be given the resources to conduct independent technical investigations into failures and accidents (Appendix 15)."

These recommendations will be considered further in Chapters 21 and 22. A note of dissent was attached to the report by 2 members of the Committee who argued that as a matter of principle the responsibility for occupational health and safety in any industry should not be held by the department with policy responsibility for that industry; and that if there was to be a single agency for offshore safety it should be part of the HSE.

16.14   The Government's reply to these recommendations, in a statement deposited in the libraries of both Houses of Parliament on 3 November 1980 was as follows:

"Accepted in principle.

The Prime Minister has decided that the Secretary of State for Energy should take over the present responsibility of the Secretary of State for Employment for occupational health and safety offshore under the provisions of the Health and Safety at Work etc Act 1974. This means that the Secretary of State for Energy will in future carry sole Ministerial responsibility for all aspects of offshore safety, save for the responsibility for the safety of ships and seafarers engaged in offshore work, which will remain, as the report recommended, with the Secretary of State for Trade. In discharging this responsibility, the Secretary of State for Energy will be advised on policy matters (including the need for any new legislation) by the Health and Safety Commission (HSC), who will in turn look to the Petroleum Engineering Division (PED) of the Department of Energy for advice. HSC's role will be extended to include advice on structural safety and safeguards against fires, blowouts and other operating emergencies offshore, (on which advice has previously been given to Ministers direct by PED).

In the case of diving safety, PED and the Health and Safety Executive (HSE) have worked closely together in the production of Unified Diving Regulations which will soon be issued. Advice to the HSC on diving matters will continue to be given jointly by PED and HSE.

PED will continue to enforce the requirements of the Mineral Workings (Offshore Installations) Act 1971 and the Petroleum and Submarine Pipe-lines Act 1975, and to act as the agents of the HSE in enforcing offshore the requirements of the Health and Safety at Work etc Act 1974. Responsibility for enforcing the HSW Act in connection with pipeline works will be transferred from HM Factory Inspectorate to PED.

PED will be strengthened by the transfer of Inspectors from HSE both for policy and enforcement work on occupational health and safety offshore.

The Government believes that these arrangements will enable the development of offshore safety policy and the enforcement of safety standards to be developed in the most efficient and effective way."

16.15   Following this reply a new agency agreement was made with effect from 1 June 1981. This was implemented by arrangements outlined in correspondence between the HSE and the PED, in particular in a letter dated 23 March 1982. In addition to providing for the continued enforcement of the HSWA and its regulations on behalf of the HSE it outlined arrangements for (i) advice from the PED to the HSC and the HSE on diving operations and the safety of workers on installations; (ii) the development of health and safety regulations for submission to the HSE and the HSC; (iii) the appointment and training of inspectors; and (iv) the presentation of annual programmes of, and reports upon, work undertaken under the agency agreement. The agency agreement provided a channel for the exchange of information and advice between the HSE and the PED, principally at the instance of the latter. The HSE provided training courses, workshops and the secondment of personnel.

The organisation and functions of the Safety Directorate of the PED

16.16   Mr Priddle stated that in the light of the legislative framework the DEn saw its role as "developing appropriate standards and controls within that framework; monitoring and enforcement of compliance with the legal requirements; promoting the interests of safety generally through the development and dissemination of information and advice". Much of this work is carried out by the Safety Directorate of the PED which was not formed until 1987, under Mr Petrie as its first Director of Safety. Prior to that time various safety functions were performed in 5 out of the 6 branches into which the PED was then divided.

16.17　The work of the Safety Directorate (see Fig 15.1) is performed by the following branches:

(i)　the Installations and Well Engineering Branch. This specialist branch is concerned with (a) consents for exploration and appraisal wells offshore; all onshore wells; and offshore development wells; and (b) the safety of offshore structures and their equipment through the development of guidance for certifying authorities and the DoT, and monitoring the work of those bodies. It also handles the registration of installations, the issue of well control certificates and considers regulations and guidance for abandonment operations.

(ii)　the Inspections and Operations Branch. This specialist branch is concerned with the health, safety and welfare of offshore workers and the protection of the marine environment. The work of inspectors from the Aberdeen and London offices has already been mentioned in Chapter 15. Further sections deal with general issues of occupational health and safety, the prevention of pollution, security of installations and contingency planning. This branch also maintains liaison with the Marine Directorate of the DoT in regard to life-saving appliances and other maritime matters such as standby vessels.

(iii)　the Pipelines Branch. This specialist branch is responsible for authorising the construction of, and enforcing safety regulations in respect of, pipelines under the PSPA. It also fulfils a similar function in respect of onshore pipelines on behalf of the HSE under a separate agency agreement. Enforcement involves monitoring the implementation of quality assurance programmes and undertaking a technical evaluation of a sample of key areas.

(iv)　the Diving Branch. This support branch seeks to promote the health, safety and welfare of divers in relation to installations and pipelines.

(v)　the Research and Development Branch. This support branch commissions and manages all research for other branches. The current research programme involves an expenditure by the Department of Energy of about £6m per annum.

(vi)　the Safety Policy Branch. This branch provides assistance to the Director in developing safety policy and strategy; carrying out an on-going review of legislation; preparing the annual plan and report for submission to the HSE; and administering requests for exemption from regulations under the MWA.

16.18　From the above it can be seen that the Safety Directorate performs a number of functions in addition to that of monitoring and enforcing compliance by operators with the relevant legislation. It audits and provides guidance to the certifying authorities and the DoT. It grants consents in respect of wells and authorises the construction and use of pipelines. It should also be added that it is consulted about, and can express reservation on safety grounds in regard to, stages in the licensing system, namely (i) the issue of licences; (ii) the approval of operators; and (iii) the approval of development plans at the "Annex B" stage.

16.19　The Safety Directorate employs about 45 specialist staff. If those employed by the certifying authorities and the DoT are included the total would be about 300. As part of a "devolved system" the Director of Safety is responsible for all policy issues within his field of activity. The other part of the PED is the Exploration Appraisal and Development Unit (EADU) which maintains control over exploitation of resources in the UKCS. The PED is one of 8 divisions reporting to the Deputy Secretary. Central management issues are decided by a management board on which he sits along with the Principal Establishment and Finance Officer under the chairmanship of the Permanent Under-Secretary. This board covers the allocation of funds to the various divisions. Mr Priddle said that by virtue of his wider responsibilities and his contacts with persons at a more senior level in the outside world he would bring a wider experience to bear on the development of policy and would intervene as and when he felt it necessary, for example in his discussions with Mr Petrie as to the current functioning of the Safety Directorate. Mr Petrie himself was in a position of being able to discuss any matter of concern directly with Ministers should he think this necessary. Mr Priddle also said that he regularly discussed matters of policy with the head of the PED.

Links between the PED and the HSC and the HSE

16.20   As stated above the PED is responsible to the HSC for the enforcement of the HSWA and its regulations offshore and for providing advice for offshore safety in general. The HSE is not responsible for health and safety policy offshore, but the HSC is. Each year the PED is expected to submit to the HSC for their approval a programme of work for the next financial year and an outline forward programme for the next 5 financial years. It is also expected to submit each year to the HSC a report outlining their activities in the past year and following the outline previously approved by the HSC. Neither the HSC nor the HSE carry out a detailed audit of the work of the PED. In accordance with Government accounting conventions the PED remains responsible for the efficiency and economy with which it implements its responsibilities, such as under the agency agreement.

16.21   The PED also acts as one of the advisers and assessors to the OIAC, which is composed of representatives of the TUC and the CBI and provides advice to the HSC in relation to the whole oil and gas industry both onshore and offshore. This committee is chaired by the HSE. Both the HSE and the Marine Directorate of the DoT may also send assessors and advisers. This committee is on occasion used as an alternative source of guidance to the Safety Directorate. The Director of Safety is able to participate in the discussions of the management board of the HSE. He presents the proposals of the Safety Directorate to the HSE before they are submitted to the HSC and participates in consideration of any other proposals which may have relevance to the offshore petroleum industry. Apart from these official points of contact DEn inspectors are members of the HSE's technical working groups and industry liaison committees. Further, staff of the Safety Directorate participate in workshops presented by specialist groups within the HSE.

16.22   The HSE has always been ready to assist the PED in any material respect. It has provided training and technological support. It has also seconded inspectors from time to time.

The development of regulations and guidance

16.23   Before a regulation is made under the MWA or the PSPA the Secretary of State for Energy is under a statutory duty to consult with organisations in the United Kingdom appearing to him to be representative of those persons who will be affected by it. In practice he also takes the advice of the HSC. Where the HSE proposes a set of regulations under the HSWA the PED is asked to advise the HSC whether they should be applied offshore. The PED has also provided advice to the HSC as to the use which should be made of the various Acts to which reference has been made earlier in this chapter. I will discuss this subject in more detail in Chapters 21 and 22.

16.24   In recognition of the need for more specific guidance as to the implementation of regulations the PED have produced guidance notes on broad areas such as design and construction (relevant to the Construction and Survey Regulations); on life-saving appliance and fire-fighting equipment (relevant to the regulations bearing these titles); and in regard to training. These notes are subject to formal consultation when issued or amended. Guidance is also provided on more specific matters in the form of diving safety memoranda, continental shelf operations notices, safety letters, safety notices and safety alerts. As regards guidance notes and safety notices Mr Petrie stated that the DEn expected operators to take account of their substance "much as we would expect them to take account of other similarly authoritative codes and standards issued by standards-making bodies such as the British Standards Institution and the professional bodies. In our experience they do take account of them."

Certifying authorities

16.25　For the purposes of Sec 3 of the MWA the Construction and Survey Regulations require that both fixed and mobile installations should be the subject of a certificate of fitness issued by a certifying authority which is valid for a maximum of 5 years. Although the Secretary of State for Energy may act as a certifying authority in practice this role is undertaken by one or other of 6 bodies which were appointed by him for the purpose and which work under contract with the operators who bear the entire cost of their services. This is one of a number of respects in which the regime was originally modelled on that developed for ships. The bodies are essentially ship classification societies. It is clear that there were seen to be practical advantages in drawing upon their expertise with particular reference to the structural and marine aspects of installations. The role of certifying authorities in the regime was clearly endorsed by the Burgoyne Committee (paras 6.8-12 of their report).

16.26　The work of a certifying authority may be put briefly under the following broad heads: (i) the assessment of design and method of construction; (ii) the assessment of the operations manual; (iii) the inspection of the installation and its equipment in fulfilment of the requirements for major surveys over a 5 year period and for annual surveys. Events such as damage and structural deterioration; and alterations, repairs and replacements require to be reported to the certifying authority with a view to its determining whether or not an additional survey is required. It also reviews the operator's programme of planned maintenance under the Operational Safety, Health and Welfare Regulations. The basic requirement which requires to be fulfilled by the above work is that the design and construction should comply and continue to comply with the Second Schedule to the Construction and Survey Regulations; and that the operations manual should contain information, guidance and instructions which are adequate and appropriate in relation to the installation. The Second Schedule covers environmental considerations, foundations, primary structure, secondary structure and fittings, materials, construction and equipment. In the light of its findings a certifying authority may attach a limitation or qualification to a certificate of fitness.

16.27　Much of the Second Schedule is concerned with structural requirements directed to enabling the installation to withstand the environmental and other forces imposed upon it. However, it also contains some requirements which are of significance for the prevention and mitigation of incidents on installations. These include those relating to: (i) equipment; (ii) material; (iii) living accommodation; decks, stairways, etc; and escape routes; (iv) ventilation heating and cooling; lighting; and emergency power supply. The work of a certifying authority is limited by the scope of the Construction and Survey Regulations. Within that scope it appears to be concerned with the conceptual and detailed design of the structure and the operation of the platform as a marine installation. As regards process plant, the design concept would be taken into account in deciding whether a particular item of equipment was "suitable for its intended purpose". According to Mr Petrie the certifying authority would consider the operating parameters of a proposed system and assess whether it could safely operate within them and what controls were provided to limit them. However, the certifying authority is not, in general, required to undertake a conceptual analysis. In particular it is not required to review plant design in relation to major hazards. It should be added that a body which was a certifying authority could undertake a conceptual design analysis on a consultancy basis so long as it was not in conflict with its role as certifying authority in the particular case.

16.28　As I have mentioned above the PED provides guidance notes in supplement to the requirements of the Construction and Survey Regulations. These are based on recognised standards and procedures which have been established by internationally recognised organisations. The fourth edition of these guidance notes was prepared during the currency of the Inquiry. It states *inter alia* that:

> "The certifying authority has discretion to accept methods, techniques, standards and codes of practice other than those in Guidance subject to being satisfied that

the installation substantially complies with the requirements specified in Schedule 2 to the regulations and that no diminution in safety or integrity will result. The DEn should be informed of any proposal of unusual or controversial character."

This statement is broadly in line with the text of previous editions. Although a large measure of discretion is entrusted to the certifying authorities only the DEn can revoke a certificate of fitness or can take action to prevent an installation being operated in the absence of a valid certificate of fitness. The DEn does not itself undertake assessment, survey or certification. Its auditing of the work of certifying authorities (as from 1987) is intended to confirm that its requirements are being complied with. Audits are carried out both on a random basis and on the basis of a specific project which is of interest to the Department.

16.29 Lloyd's Register of Shipping acts as the certifying authority in respect of over 80% of the fixed installations in the UKCS. At its London headquarters it approves plans and appraises designs: outport offices such as in Aberdeen, carry out the work of surveying, subject to procedures, instructions and specialist support provided from headquarters.

The Department of Transport (DoT)

16.30 So far as installations are concerned the Marine Directorate of the DoT (formerly of the Department of Trade) carry on work on behalf of the DEn in relation to fire-fighting equipment and life-saving appliances. Fire-fighting equipment for new installations is examined by the DoT in order to see that it is in accordance with the plans, the Fire Fighting Equipment Regulations and the relative guidance notes. It is stated in the guidance notes that the DoT has discretion to accept methods and techniques equivalent to those outlined in the guidance notes subject to being satisfied that no diminution of safety will be involved: but that the DEn will be informed if the proposal is of an unusual or controversial character. Fire-fighting equipment is subject to examination by the DoT every 2 years in order to see that it is properly maintained and in good working order. Life-saving appliances for new installations are examined by the DoT in order to see that they are in accordance with the Life-saving Appliances Regulations and the relative guidance notes. Their approval of the appliances is required. The appliances are subject to examination by the DoT every 2 years in order to see that they are properly maintained and in good working order. These arrangements provide another instance where the expertise of those familiar with the regime which applies to ships has been used as part of the regime for installations. The work of the DoT on behalf of the DEn is subject to audit. The DoT are also concerned with navigational aids; and with whether standby vessels provided under the Emergency Procedure Regulations meet the code for assessment of their suitability in accordance with a voluntary agreement with UKOOA.

The Civil Aviation Authority

16.31 The other body which has a specific responsibility in regard to installations is the CAA which is responsible for safety in commercial helicopter operations.

The UK onshore safety regime

16.32 The origins of the HSC and the HSE have been briefly referred to in paragraph 16.8 above. Their general aims can be clearly understood by reference to the following passages in the HSC's published plan of work for 1989-90 and beyond:

"We and the executive are regulatory bodies, concerned with protecting people from harm. This is true in the formal sense of our having a statutory duty to submit proposals for regulation and the executive having a similar duty to make arrangements to enforce them. It is equally or more true in the profounder sense of our being the prime movers in a vast activity, undertaken day by day within industry, to prevent accidents and ill-health and to protect workers and the public, essentially from the release of the energies that work involves. The first of the statutory duties specified to us in the HSWA is to 'assist and encourage' those engaged in this task." (Para 26).

"Our basic aims continue to be to:

(a) stimulate and guide the efforts of industry to achieve higher standards of health and safety at a cost that is realistic; and

(b) protect both people at work and the public who may be affected by risks arising from work activities, and keep them properly informed about the risks and the protective measures adopted." (Para 65).

The Health and Safety Commission

16.33 The HSC is composed of 8 members nominated by the CBI, the TUC and local authority associations, and a chairman appointed by the Secretary of State for Employment. It proceeds essentially on the basis of consensus; and, according to Mr Rimington, takes every possible step to ensure that those affected by its activities are in at least broad agreement. It makes substantial use of advisory committees, either in regard to particular types of hazard or (as in the case of the OIAC) in regard to whole industries. In these committees much of the work of determining safety standards is carried out.

The Health and Safety Executive

16.34 The HSE is a corporate body of 3 persons appointed by the HSC, namely the Director General, the Deputy Director General and the Director of the Safety Policy Division. It has about 3,600 employees, mainly inspectors and technical, scientific and medical experts. Its management board includes the Chief Inspectors of the various inspectorates which are concerned with the enforcement of industrial safety and health. The HSE is the licensing authority for nuclear installations. Following recent changes the Nuclear Inspectorate and 4 divisions which are concerned with policy and planning report to the Director General. The Mines Inspectorate, Field Operations (which includes the Factory, Agricultural and Quarries Inspectorates and the field consultancy groups which provide technical and scientific support to the inspectorates) and the divisions concerned with technology, research and laboratories report to the Deputy Director General (see Fig 16.1).

16.35 Para 164 of the plan already referred to describes the expertise of the HSE as made up of 3 main kinds, namely (i) policy branches which advise the HSC on matters such as possible changes in legislation or standards; (ii) inspectorates which secure compliance with legal requirements and accepted standards through inspection, advice, investigation of incidents and, where necessary, enforcement; and (iii) the technical, scientific and medical group which are principally responsible for promoting and supplying excellence in the scientific and technical advice available to the HSC and others concerned with safety. The plan also states:

"Our function is to oversee almost all aspects of industrial safety and health in the UK, whether they affect people at work or the public. We lay down the standards for the safe conduct of virtually all industrial processes and the safe use and transportation of dangerous materials and pathogenic organisms, frequently following international negotiation. In no other country is this very large task so largely concentrated in a single body; in most it is distributed over many central and regional bodies. This concentration in the hands of the Executive of a wide range ..of professional and scientific expertise produces advantages beyond mere economies of scale. Our responsibility in relation to a wide field of risks enables fruitful exchanges of experience and ideas. It qualifies us to speak with authority on general questions concerning the nature and tolerability of risks, necessary and at the same time acceptable controls, and effective approaches to enforcement; and gives us advantages internationally. But we can only carry out our work through new

264

HSE Organisation Chart

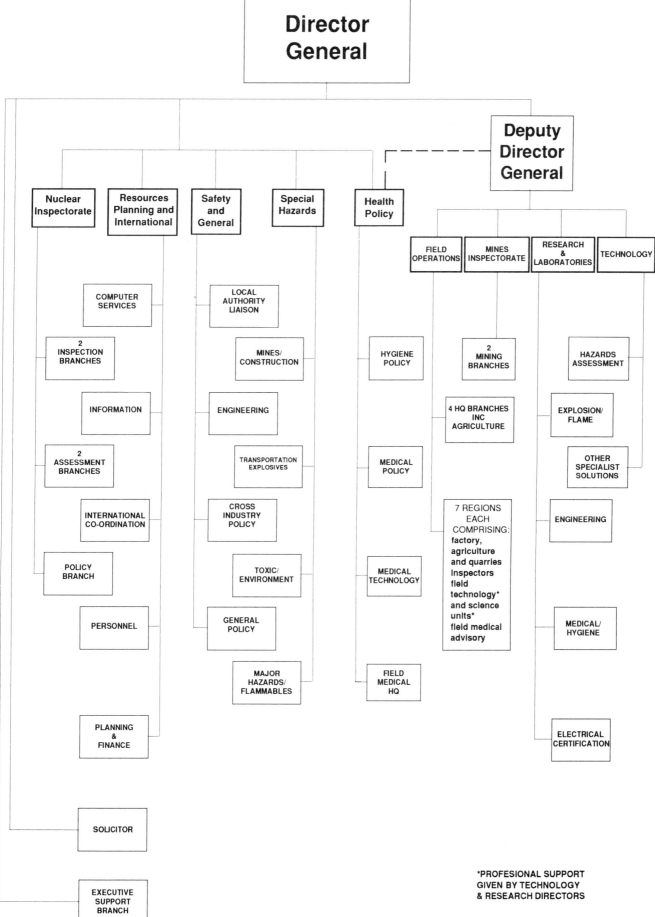

265

Fig. 16.1 Simplified organisational structure of the Health and Safety Executive.

measures, getting people together to decide on best practice, and by stimulating and encouraging those within industry who carry the legal responsibility for health and safety in individual enterprises." (Paras 66-67).

Apart from its own resources the HSC through its research committee commissions a large amount of extramural research. This includes work undertaken by and in conjunction with the nuclear industry.

The HSC's approach to regulations

16.36   The progressive replacement of the existing onshore legislation relating to health and safety has necessarily required a long period for its implementation. The plan already referred to states that by then HSC had brought about the repeal of 143 sets of regulations and introduced 35 "new packages on modern principles". It also stated that the law on a high proportion of hazards, including that on hazardous and dangerous substances and their transportation, had been comprehensively reformed, and major packages on mining, electricity and pressure systems were progressing well (para 32). The new style of regulations specify principles rather than solutions and are thus intended to encourage innovation on the one hand but be effective against lack of precaution on the other. Mr Rimington explained that one implication of the revised approach was that regulations should so far as possible apply across the board ie to every industry where the hazard in question applies. However, it has sometimes been found necessary where hazards assume a special form in a particular environment for there to be special regimes. In such cases the principles of the more general regulations are applied with particular additions as appropriate. He explained that regulations under the HSWA were necessary only where in some particular respect the general requirements of the act needed to be spelt out in a form which was in some sense mandatory, such as the expression of a legal duty, a principle of action or a strict requirement of some detailed kind. The new style of regulations were backed up by approved codes of practice which created a presumption that the law had been breached unless it could be shown that some equally satisfactory approach had been adopted; or by guidance; or by both. The fact that codes of practice and guidance were neither mandatory nor required complex parliamentary procedures for their amendment made them an ideal vehicle for incorporating the results of changing technology.

Compliance with legislation

16.37   Mr Rimington described the primary object of inspection as being "to stimulate the operator to carry out his duty to maintain safety, against the background that the inspector can and will apply coercion through the courts where this is necessary or salutary". The intention was always to ensure that a high standard of compliance with reasonable standards had been attained. He also said: "An inspector's immediate purpose in visiting is to satisfy her or himself that systems exist that are likely to lead to the identification and prevention by management of significant faults, and that the attitude of management is conducive to this." It was also very important that inspection should be targeted on aspects that were critical to safety "so that time is not wasted and discoveries, if made, are likely to open up further vistas for enquiry". In this context he drew attention to the fact that the HSE's inspectorates were of two kinds, namely those which concentrated on particular industries where the inspections covered plant which was likely to be relatively familiar to the inspectors who would generally be recruited from highly qualified people with experience of the industry itself; and those inspectorates (particularly factories and agriculture) which cover much larger territories. The latter were recruited at graduate level. They often though not necessarily possessed technical degrees or qualifications. All received very substantial initial training. Many inspectors spent a large part of their working lives dealing with particular industries.

16.38   Inspectors in the HSE are able to call upon a wide range of resouces within the HSE's Technology Division, the Employment Medical Advisory Service and the specialist sections of the Nuclear and Mines Inspectorates; and the forensic and general scientific capabilities of the Research and Laboratory Services Division. The specialist inspectors and scientists when called upon in this way did not lead an inspection or investigation but were called in as experts to investigate particular aspects and also as necessary to give expert evidence in court. The specialist inspectors in the Technology Divison and the Nuclear and Mines Inspectorates generally have substantial industrial experience. They and the HSE's scientific staff are in contact with the latest advances in thinking and good practice. Specialist inspectors are also engaged in the assessment of Safety Cases under the Control of Industrial Major Accident Hazards (CIMAH) Regulations, whether in the nuclear or non-nuclear areas; and expertise is maintained in the techniques involved in the assessment of risk.

16.39   Particular mention should be made of the HSE's Accident Prevention Advisory Unit (APAU). This unit has been maintained since 1977 in order to exercise influence on the management of safety and the design of safe systems. Its main task has been to carry out safety audits in co-operation with large companies and undertakings in order to investigate the standards of management control and advise how a structured safety system at corporate and subsidiary levels can contribute to high standards of health and safety. In 1984 the HSE had put forward a scheme for "safety assurance", linking safety and quality assurance. Under this scheme the HSE would have audited the employers' safety systems. However, this was on the basis that the employers would be exempt from basic inspection. The proposal was dropped after it did not prove widely attractive to employers and was strongly opposed by trade unions. Mr Rimington said that the APAU had gradually gained for itself a considerable reputation and its experience had enabled it to begin to formulate a body of knowledge and principles on safety management. Monitoring of companies which had accepted the advice of the APAU had shown conclusively that lasting results had been obtained not simply on the basis of the advice given but through stimulating the attention of management to a subject which had been frequently neglected or regarded as technical or obscure. The HSC's plan of work for 1989-90 and beyond which has been referred to earlier in this chapter records at para 98 that the unit evaluates safety monitoring packages and advises the HSE's inspectors and the public about their use. Field inspectors would take suitable opportunities to inform companies about them. The unit also intended to evaluate the potential link between various quality concepts such as that in BS 5750 - the UK national standard for quality systems - and standards of safety. They were also collaborating with companies and other parts of the HSE in an attempt to define the costs of occupational accidents and ill-health and quantify the economic benefits derived from high standards of occupational health and safety.

The Norwegian offshore safety regime

16.40   The exploitation of petroleum resources in the NCS is controlled by the issuing of licences and the granting of consents which enable licensees to progress through various stages leading to production. Since 1979 the Norwegian Petroleum Directorate (NPD) (which was established in 1973) has been responsible to the Ministry of Local Government and Labour in matters relating to the working environment, safety and emergency preparedness; and to the Ministry of Petroleum and Industry for the administration of petroleum resources in an efficient manner. On this basis the NPD's objectives are "actively to contribute to a sound administration of the Norwegian petroleum resources through a balanced evaluation of the natural, safety-related, technological and economic aspects of the activity within an overall social framework." The control of Statoil which engages in the business of petroleum production and is wholly owned by the Norwegian state is exercised directly by the Ministry of Petroleum and Industry. In its supervisory capacity the NPD has the task of seeing that both safety legislation and the terms of licences, consents and approvals are complied with. Under authority delegated by the Ministry of Local Government and Labour the NPD has the power to issue regulations and conduct overall safety evaluations. In exercising its supervisory authority the NPD obtains assistance from other public bodies, institutions and companies with special expertise. The NPD also acts as adviser to the 2 Ministries, and is responsible for providing guidance to all participants in the petroleum industry.

The organisation of the NPD

16.41 The highest administrative authority of the NPD is the board of directors which consists of the chairman and 7 members. The day to day business of the NPD is in the hands of the Director General. Apart from the administration and legal branches and the information office, the NPD is divided into 2 divisions, namely the Resource Management Division and the Safety and Working Environment Division, the work of which respectively relates to the 2 responsibilities of the NPD referred to in para 16.40.

16.42 The main objective of the Safety and Working Environment Division is to establish, maintain and further develop a fully satisfactory level of safety and working environment within the petroleum industry. It consists of the following branches:

(i) the Supervisory Activities Branch. Its main task is to manage supervisory activities in the Norwegian part of the continental shelf (NCS). Six Heads of Supervisory Activities are responsible for the supervision of specific operators. In the NCS there are about 40-50 installations, operated by 12-15 companies, including many which operate internationally. About 15,000 persons work offshore, of whom about 5000 are there at any given time.

(ii) the Technical and Working Environment Branch. This provides a pool of expertise in safety and the working environment. It is responsible for providing personnel on a priority basis for various tasks such as the development of regulations.

(iii) the Strategy Branch. Its principal responsibilities are that of offering advice and guidance, undertaking development of regulations and co-ordinating and managing certain tasks. In addition it is responsible for the execution of a number of administrative functions for the division as a whole.

About 90 professional staff are employed in the Safety and Working Environment Division. Mr Ognedal stated that installations in the NCS could be visited 3-5 times a year; and that about 80 of his staff would be expected to be a couple of times offshore each year. This could range between a simple verification taking a few hours and a week's investigation.

16.43 The purpose of the Resource Management Division is to survey petroleum resources and evaluate alternative ways of extracting and utilising these resources in the best way, with a view to advising the Ministry of Petroleum and Energy how to control exploitation for the greatest benefit of society.

The NPD's approach to legislation

16.44 Since 1985 the NPD has been responsible for the review and reformulation of regulations which prior to that time had been issued by 9 independent Government agencies which had separate responsibilities in different areas of the petroleum industry. Since the beginning of 1987 the NPD has been examining all existing regulations with a view to reformulating them and reducing their number by about 50%. Mr Ogendal said that about 15% of the total resources of his division were involved in this work. The new style of regulations were concerned with objectives to be achieved and so were "goal-setting". In connection with the development of new regulations the NPD had created a forum of some 35 persons drawn from the NPD, the industry, the Government and trade unions. A reference group had the oversight of the development of the regulations. In addition experts had been chosen for their particular expertise in relation to the individual regulations to be drafted. Mr Ognedal said that this forum enabled the NPD to exchange ideas with the industry from a very early stage. The NPD also intended to provide supplementary documentation in the form of guidance notes. The available codes and standards would play an important part in the future Norwegian legislation.

Internal control and supervisory activities

16.45 A fundamental principle of the Norwegian safety regime is that since the operator controls his business he therefore should control the safety aspect of his operations. This principle began to be developed in 1975 and is currently formulated in Internal Control regulations stipulated by Royal Decree of 28 June 1985 in pursuance of a number of Acts of the Norwegian Parliament concerned with safety in petroleum activities, worker protection and the working environment and protection against pollution in petroleum activities. In accordance with the regulations the licensee through internal control is to ensure that the activity is in accordance with the provisions stipulated in and in accordance with those Acts (Reg 3). "Internal control" is defined as: "All systematic actions which the Licensee shall initiate to ensure that the activity is planned, organised, executed and maintained according to requirements stipulated in or in accordance with acts or regulations." (Reg 2). The regulations set out a large number of matters which the licensee's internal control is to ensure, such as "that safety evaluations are undertaken both prior to start of exploration drilling, prior to the final selection of project plan and its subsequent phases, including the operating phase"; "that the licensee's and contractors' employees are given necessary training"; and "that the contractors' Quality Assurance system is evaluated and assessed, and is subject to system audit". According to the regulations responsibility for the monitoring and enforcement of the control system is to be assigned to a separate unit within the licensee's organisation. This unit is to have sufficient organisational freedom to monitor and enforce all subordinate control systems and to perform system audits on these. This organisational unit is normally to be placed outside the operative responsibility. The licensee is to arrange the organisation in such a way that the unit normally reports to a higher organisational level than those units which it is to monitor. A "system audit" means a "planned and systematic examination of systems to ensure that these have been established, followed and maintained as specified". The internal control system is to be kept up to date in a systematic and controlled manner; and the up-dated information is to be communicated to the NPD and within the licensee's organisation, workforce and contractors.

16.46 In order to ensure compliance with the requirement for internal control the licensee has accordingly to establish and describe the system which is used to control its own activity. This description is the main tool which the NPD use in carrying out its supervisory activities which are separate from and additional to the internal control exercised by the licensee. The object of NPD's supervisory activities is to make a systematic assessment of the internal control system of the company in order to check that its activity is correctly reflected in the documentation and is performed within the requirements of the law. The NPD performs this assessment in 2 ways - by audit of the operator's systems and by verification of the output of those systems. The NPD endeavour to schedule an audit of each licensee at least once every 3 years; normally it is more frequent than that. The NPD select a section of the licensee's operation and audit this section from the most senior person down the management structure. In advance of the audit the company is advised in general terms of NPD's plan for the following year; and will be required to provide the NPD with specified information and documentation including the company's own plan for auditing. Having considered this documentation the NPD plan one or more audits and the timescale for them. An audit team, usually consisting of 4 or 5 persons under a team leader plan a questionnaire for use in interviewing company personnel. Following the interviews, which are informal in character, the team then may carry out a verification on the installation in order to ascertain whether a particular procedure has been set up, is documented and is understood and operated by the appropriate personnel. Equipment may also be checked. The NPD may involve consultants in carrying out specific checks, for example in the areas of welding, corrosion and diving. Sometimes the NPD use management consultants in the planning of the audit. The company which is being audited is invited to have a representative present as an observer throughout the audit. The elected safety delegate from the workforce also normally takes part, especially offshore. When the audit has been concluded the NPD make a presentation of the result to the company following the presentation by the company's observer of his findings. Thereafter the NPD prepare a report and send it to the company for their reaction. The final report will be sent to the company with a covering letter detailing the points which have been identified in the audit and require action by the company. The letter will also ask the company to submit to the NPD a plan for corrective action by a given date. Both the final letter and all attachments to it are available both to the press and to the public. Within the NPD the responsibility for following up its findings rests with the appropriate Head of Supervisory Activity. The Heads are in daily contact with the companies for which they are responsible. The Ministry of Local Government and Labour will not normally become involved but may do so if something very serious is thrown up by the audit. The Ministry do not receive a copy of the NPD reports as a matter of routine but are aware of the companies which the NPD are auditing as this information is contained in the NPD Supervisory Activities Plan for each year about which the Ministry is informed. The supervisory activity of the NPD is paid for by the company which is being audited. A major audit will take on average 3 weeks from the start of detail planning to the writing of the report. In the planning of an audit a decision requires to be taken at an early stage as to whether any contractors should be included in the activities. This will depend on the extent of auditing by the operator. It may be necessary to involve contractors either in the NPD audit or at the stage of verification. Mr Ognedal stressed that it was necessary for the objectives and scope of the particular audit to be described clearly. The methods of obtaining verification by reviewing documents and carrying out inspections require to be thoroughly discussed and decided with the objectives of the particular audit in mind. A knowledge of the available documentation was essential. It was also important for the necessary expertise for the particular audit to be discussed and decided on in advance. He saw this system, which was based on open communication between the company and the NPD, as presenting a new challenge for both.

16.47　This regime has never used a separate body to carry out the assessment of design or surveys with a view to certification of installations. Mr Ognedal mentioned 2 reasons: the insistence on a one point responsibility and the aim of having short lines of communication between those who supervised and those who drew up the regulations. It was and is a matter for individual companies to decide whether they wished to use the expertise of such bodies in their internal control activities. Further the NPD does not certify or approve company systems or procedures. The point at which they exercise control is when application is made by a company for the required consent on the part of the NPD to a stage in their activities. At that stage the NPD can refuse consent if they are not satisfied with the particular application. They can also take into account any matters which are outstanding, such as the failure of a company to present its evacuation plan to the NPD. The NPD can refuse consent until the plan is forthcoming.

16.48　The NPD is provided with powers to enforce the legislation. These range from the imposition of day fines to the shutting down of a company's activities. An NPD inspector has the power to shut down an activity on the spot. He would immediately inform the NPD at Stavanger where a decision would be taken on the information available as to whether the shutdown should be continued or not. Mr Ognedal stated that if it came to his attention over a period of time that a company was not performing adequately he could summon the company's "top man" to see him at very short notice. He had found this to be very effective in producing results. Where a serious problem had been identified in relation to a company the NPD could recommend to the Minister that the company be required to see the Ministry or the Minister himself; or that the company was not fit to continue operations.

16.49　Mr Ognedal estimated that almost half of the work of his division was devoted to supervisory activities covering present and new projects and the evaluation of new applicants for licences. About 10% of its time was concerned with giving information and advice. So far as could be done his division produced a plan of activities for the forthcoming year. In 1990 it was intended to obtain further insight as to the competence required of personnel in the NPD, its consultants and other agencies with which the NPD worked. This was intended to provide specific guidance as to future training and assist in recruiting philosophy. The priorities for 1990 were to ensure better compliance with the Act relating to the working environment; to ensure activity in the very early stages of any project; and to follow up a re-evaluation of the older installations.

16.50   Personnel employed by the NPD are recruited both from the offshore industry and from the shipping and engineering industries. Training needs are defined by the manager and the employee and a plan is created. Training comprises on-the-job training together with courses on the legal framework and the philosophy of internal control. A new recruit could also be taken into an audit team as part of his training. Personnel are also seconded to the companies for a period of time. Most recruits have university training and more than 60% are graduate engineers. The NPD also recruit university graduates with degrees in other subjects such as management and social sciences. Mr Ognedal would not expect a new recruit to perform an audit or carry out verification activities on his own for at least 2 years after he joined the NPD. However, he stressed that it was the procedure of the NPD for at least 2 persons to go offshore for a supervisory activity. When a new recruit began work a member of staff would be appointed to have special responsibility for him and his training on behalf of the manager. Although personnel could be lost to industry the turnover was only about 7%, so the numbers were "normally pretty close to maximum".

Future trends in offshore developments

16.51   In approaching the subjects discussed in later chapters, and in particular Chapters 19 and 20 it is right that I should bear in mind what the Inquiry heard as to the future trends in offshore developments since these may give some indication of how frequently a certain type of feature which is of interest is likely to be installed. One example is the inter-connection of installations by pipeline. A study of future trends may also assist me in determining whether and how far it is advisable or practicable to make a recommendation. On this point the diversity of future types of development may be of some significance.

Trends in the size of fields

16.52   Although the number of oilfields under development increased between 1980 and 1988 their average size decreased over that period. To illustrate that, Dr Taylor pointed out that the 15 oilfields in production at the end of 1980 had an average size of 855 million barrels; whereas the average size of the 26 oilfields brought into production since then was 173 million barrels. A survey carried out by UKOOA in mid-1988 had shown that there were then 44 oilfields in production or under development which had an average size of 323 million barrels; whereas there were 92 oilfields which had been discovered but were not yet developed having an average size of 52 million barrels. Accordingly the undeveloped oil discoveries averaged only one sixth of the size of the fields already in production or under development. Oil production from existing fields which was currently a little above 2 million barrels per day was in decline and was projected to halve by the mid-1990s. As at mid-1988 there were 25 gas-fields in production or under development and 35 discovered but undeveloped. Gas production which was currently about 4500 million cubic feet per day was projected to halve by the end of the century. This decline in oil and gas production would be slowed down by production from new fields some of which had already been found and others yet to be discovered.

Factors affecting the pace of development

16.53   Dr Taylor said that since 1984 a number of changes had occurred which had significantly influenced the outlook for the upstream petroleum industry. For example oil and gas prices had fallen substantially; the cost of new developments had been significantly reduced, in particular as the result of the use of heavy lift barges; and changes in depreciation allowances for corporation tax had reduced the profitability which was achievable on new investments. He said that in assessing potential future production, whether generated from existing but as yet undeveloped discoveries, or from discoveries yet to be made, the rate of appraisal and development of the inventory of discoveries depended upon a price and fiscal environment which supported commercial development, the availability of cost-effective technology and the provision of adequate resources to support the activity.

The types of development

16.54   The earliest installations in the North Sea were in the southern area (all for gas) where the water depth was relatively shallow. Later developments were in the central and northern areas in deeper water where a number of major oil discoveries were made in the 1970s. Larger and more substantial installations were required for the deeper water areas. In these areas drilling, processing, utilities and accommodation were integrated into a single platform.

16.55   There are considerable differences in the environment in different parts of the North Sea. Dr Taylor illustrated this by reference to the range of depth of water, wind speeds and wave heights. Because of these differences and other factors such as the geology and geography of the reservoirs there were many different types of offshore installation. In the southern area there were over 100 small fixed platforms mostly fairly close together. In one part of this area, covering about 40 miles from north to south and 50 miles from east to west there have been more than 30 gas discoveries. These included the very large Leman field which has 18 platforms spread over 15 x 5 miles. Some of the platforms consist of several structures linked together. This linking is also found in the group of "V" fields. One example was the North Valiant gas gathering installation which consisted of 4 structures linked together to make one installation. He compared these examples with the Brent area in the northern North Sea where there were 18 large and heavy platforms to cope with the deeper water and hostile environment. These structures were set out in an area which spanned 60 miles from north to south and 30 miles from east to west. The group included a number of different designs of concrete and steel platforms. Fairly close to these were the Statfjord group of 4 platforms on the median line between the UKCS and the NCS. Elsewhere in the northern area installations were more scattered. This had implications for the design with regard to the means of transporting products, the type of accommodation for personnel and the organisation of systems for evacuation of hydrocarbons.

16.56   The fields in production at the end of 1988 showed a diversity of methods of development. 31 used fixed platforms. 8 used floating production systems. 7 used subsea completions. One used the tension leg system.

16.57   Looking to the future developments for the production of oil, Dr Taylor said that fewer and fewer large fields requiring fixed structures would be discovered. The successful recovery of oil from many of the small discoveries would depend on the introduction of improved oil recovery and production systems. Heavy fixed steel and concrete platforms would increasingly be replaced by re-usable floating production systems and subsea developments. The latter were especially attractive where developments were favourably located near existing facilities into which they could be linked. Floating structures were applicable over the whole range. As regards the production of gas condensate, this would mostly be done by means of fixed platforms with some smaller developments done by subsea installations. For gas discoveries, the larger fields would normally be developed using central fixed installations tied to not normally-manned wellhead towers. Smaller gas fields would use similar unmanned towers which would be tied to existing production platforms.

16.58   Dr Taylor described a method of predictive analysis which had been carried out by UKOOA in order to examine the distribution of size and development of fields within the next 25 years. By the use of this analysis it had been estimated that out of the 101 existing and future undeveloped oil discoveries, 51 would use subsea methods, 25 would use floating production systems and 25 would use fixed platforms. About 30% of the reserves accounted for in this way would be accumulations of less than 50 million barrels. Development of the very small fields would depend heavily on the complexity of the geological structures, oil prices, the fiscal position and the available technology. Access to existing installations and pipelines would be required. Technological advances in the last 5 years had made it probable that these small discoveries would be developed by subsea completions tied into parent installations. On the basis of past experience and the current uncertain outlook facing the offshore industry the study carried out by UKOOA had adopted an average development rate of 3 oil-fields per year over the next 25 years. As regards gas-fields size was no longer considered to be of great significance in determining the order in which fields came forward for development. This had been due to the extent of the existing infrastructure of pipelines, terminals and gathering platforms which was now in place, especially in the southern North Sea. Further many oil companies had a diversity of holdings which provided them with an incentive to develop smaller fields. Against that background Dr Taylor said that it was possible to postulate a future development scenario which included 3 new oilfield developments per year (which could extend to 4 if subsea completions were involved); 6 new platforms per year for new gasfield developments; and 1 new gas condensate field development every 3 years until the end of the century and 1 per year thereafter. Dr Taylor said that it could be concluded from the overall result of the study that in the northern and central North Sea the potential existed over the next 25 years for the development of around 100-150 oilfields and 20-30 gas condensate fields requiring in total around 45-55 fixed platforms, 20-30 floating production systems and 40-60 subsea installations.

16.59   Dr Taylor referred to the installation for the Tern field as being representative of the type of manned fixed installation likely to be built in the future. The field was among the group containing between 180 and 320 million barrels of oil. The platform would be typical of the future fixed platforms required to develop the larger of the remaining small fields in the northern North Sea. Platforms of this type would be dealing with much the same operations as the larger platforms but on a reduced scale. Where satellite systems were involved oil and gas would emerge from these in single pipelines delivering to the main platform for separation. The gas would probably be used as a fuel. He illustrated the use of floating production systems by reference to the Ivanhoe/Rob Roy development. This installation was semi-submersible, kept on station by dynamic positioning. Oil and gas was collected from subsea connections through a system of flow lines and flexible risers. The development was the first to combine oil and gas production in a floating system. It would be able to pump water into the reservoir and maintain reservoir pressure. The gas produced would be taken by way of the Tartan field into the Frigg system. The oil would be transported to Claymore and then to Flotta. If necessary it could be discharged into an oil tanker near the development. Another floating system under construction was designed for extracting oil from very small reservoirs. This was described as a single well oil production system (SWOPS). It consisted of a ship-like facility which could be used as a subsea production system on an existing exploration or appraisal well and possibly to develop some of the very small discoveries I have mentioned above. The oil could be processed on board. The advantage of that system was that it could be moved elsewhere when the reservoir had been exhausted.

16.60   It appears that satellite subsea developments and small unmanned installations will increase as a proportion of future developments in the UKCS. However, the large number of existing fixed platforms and the substantial, if decreasing, proportion of new manned installations will continue to demand a high degree of attention to personal safety. At the same time technological change and the increasing proportion of unmanned installations will call for a flexible approach to regulation.

Chapter 17

Safety Assessment

17.1   I now turn to outline the specific elements in the regime which I envisage, starting in this chapter with safety assessment. I will first describe the nature and value of safety assessment (paras 17.3-7), the models for its regulatory use furnished by the Safety Case in the Control of Industrial Major Accident Hazards (CIMAH) Regulations 1984 (paras 17.8-16) and by the arrangements in the Norwegian Continental Shelf (NCS) (paras 17.17-21), and the practice and intentions of the DEn (paras 17.22-28). Next I will state my views on, and proposals for, the role of safety assessment in the regime (paras 17.29-47). I will then consider the role of quantification and quantitative risk assessment (paras 17.48-61). I will conclude by considering the implications of my proposals for regulations, for the regime and for the regulatory body (paras 17.62-71).

17.2   The Piper disaster involved the realisation of potential major hazards. There was a leak and an explosion inside a module followed by the rupture of gas risers. Although Occidental could not but be aware of the existence of such hazards, it did not possess any system which ensured that such remote, but potentially disastrous, events were subjected to systematic scrutiny. There was for major projects no comprehensive system of safety assessment and management did not appear to appreciate fully the contribution which it could make. By contrast, the evidence showed that some companies, both those operating in the UKCS and in the NCS, require the formal use of safety assessment for major projects, and did so prior to Piper. The companies which gave evidence on this were clear that these activities were beneficial in the identification and control of hazards.

Formal safety assessment

17.3   Formal safety assessment (FSA) involves the identification and assessment of hazards over the whole life cycle of a project from the initial feasibility study through the concept design study and the detail design to construction and commissioning, then to operation, and finally to decommissioning and abandonment. The techniques used include hazard and operability (HAZOP) studies; quantitative risk assessment (QRA); fault tree analysis; human factors analyses; and safety audits. The need for FSA arises because the combinations of potential hardware and human failures are so numerous that a major accident hardly ever repeats itself. A strategy for risk management must therefore address the entire spectrum of possibilities.

17.4   In accordance with the usage of the main witnesses in this passage of evidence, I shall use the term "formal safety assessment", or FSA, to mean the process of assessment and "an FSA" to mean the output from this process and in particular an assessment essentially equivalent to a Safety Case. It is with this latter that I shall be primarily concerned in the first part of this chapter.

Current use of FSA offshore

17.5   Some companies operating in the North Sea already produce FSAs for major projects. Dr M S Hogh, Manager of Projects and External Affairs, Group Safety Centre, BP International, described the formal system used within BP. There is a formal Project Review Procedure conducted at 6 distinct stages in the course of a project, starting with definition and feasibility and going through to operation, in which independent audit teams seek to identify any outstanding safety issues. There is a formal requirement to carry out HAZOP studies at the detail design stage and the results are scrutinised by the audit team. FSAs were also described by Mr R E McKee, Chairman and Managing Director of Conoco (UK) Ltd, who stated that since Piper a separate group had been created to deal with this.

17.6   The value of an FSA to the company was illustrated by the Engineering Safety Plan for the Southern Basin Gas Development, or V Fields project, described by Mr M Ferrow, Manager, Safety and Quality Assurance, Conoco (UK) Ltd. The development involved a number of gas fields some 50 miles east of the Lincolnshire coast. The exercise was modelled on the Norwegian concept safety evaluation (CSE), a form of FSA carried out at the conceptual stage. The objectives were to demonstrate the safety and reliability of the design; to detail the operational requirements and limitations; and to provide the basis for continuing safety assurance after handover. One outcome of the work was a systematic, documented review of all significant accident scenarios and the associated precautions. Another was a lead in to the operating group of all the issues which the design group felt were important for safety. Some 500 operating practices were derived from the work. The documentation comprised in its detailed form some 11 volumes. The value of the plan was demonstrated, when, following Piper, the company reviewed its safety precautions. The availability of the plan documentation made this a relatively straightforward exercise.

17.7   As his own title indicated, Mr Ferrow regarded safety and quality assurance (QA) as linked. He described FSA as a subset of quality assurance. The in-house quality assurance system was used to ensure that findings of the V Fields study just described were properly closed out.

The CIMAH model

17.8   I now turn to consider the role of an FSA in the regulatory regime. The Inquiry heard of 2 existing models for this, the onshore Safety Case and the arrangements in the NCS. The Safety Case was described by Dr A D Sefton, a factory inspector based at Sheffield and leader of the HSE's Hazardous Installations and Transport of Dangerous Substances National Interest Group.

17.9   Onshore major hazard installations are subject to the CIMAH Regulations. Reg 7 requires that the operator should provide the HSE with a written report on the safety of the installation. The report is commonly called the Safety Case. These regulations had their origins in the Flixborough disaster in 1974 and the work of the Advisory Committee on Major Hazards (ACMH) and in the Seveso disaster. They effect, and are confined to, the implementation of the EC Directive on Major Accident Hazards, the so-called "Seveso" Directive. They require demonstration of safe operation (Reg 4), notification of major accidents (Reg 5), a written report (the Safety Case) (Reg 7), updating of the report (Reg 8), an obligation to supply the HSE with further information (Reg 9), preparation of an on-site emergency plan (Reg 10) and provision of information to the public (Reg 12). There are also requirements on the local authority to prepare an off-site emergency plan (Reg 11).

17.10   The contents of the written report are specified in Schedule 6 of the regulations. The 4 main headings relate to information on every dangerous substance involved in the activity; on the installation itself; on the management system; and on the potential major accidents. The information required on the management system includes the staffing arrangements; the arrangements made to ensure that the means provided for safe operation are properly designed, constructed, tested, operated, inspected and maintained; and the arrangements for training. That required on the potential major accidents includes the potential sources of a major accident and the conditions or events which could be significant in bringing one about; the features of the plant which are significant as regards potential for a major accident or its prevention or control; the measures taken to prevent, control or minimise the consequences of a major accident; and the emergency procedures.

17.11   In the first instance the Safety Case is a means by which an operator demonstrates to itself the safety of its activities. The value of such a demonstration was illustrated in the evidence of Mr Ferrow concerning an FSA which I have already mentioned.

17.12   The Safety Case also serves as the basis for the regulation of major hazard activities, as described by Dr Sefton. Existing major hazard installations were already subject to the Health and Safety at Work etc Act (HSWA) and even before the CIMAH Regulations had been the subject of attention by the local inspector. A CIMAH site is normally visited annually. On receipt of a Safety Case the HSE first checks to ensure that all the information required is provided and to identify any matter of immediate concern. The report is then assessed by a multi-disciplinary team including specialists from HSE's Technology Division, the local area inspectors and as necessary local specialists in the Field Consultant Groups. Any matters of concern are then taken up by letter or by visit. Following this initial response, the report constitutes an important input into the inspection strategy and provides a basis for selecting areas which should receive priority attention. Examples cited by Dr Sefton were the use and maintenance of an item of hardware such as pressure relief devices or a procedural matter such as operating instructions. In the course of in-depth inspection of these items inspectors would always test management and organisation of the installation by reference to any failings detected. Many operators have reported that they found the exercise of producing a Safety Case valuable. Often it would be the first time that a report had been made of the major hazard aspects of the installation. Many stated that the exercise had led them to make changes in their approach and improvements to systems and procedures. Dr Sefton was at pains to point out that the Safety Case was not a licensing or approval system, which might be thought to transfer some of the responsibility to the licensing authority. He was not even sure that the HSE "accepted" the report. What it did was to satisfy itself that the information provided complied with that required and then use that information in its inspection; if there were any serious concerns arising out of the report, it would take them up.

17.13   With regard to the level of expertise required within the company to prepare the Safety Case, the guidance notes state:

"A partial answer is to suggest that if a manufacturer was unable to meet most if not all of the aims of the Safety Case set out in para 106 by using his own staff, doubts would arise about his competence to manage a major hazard activity ..." (para 114).

In practice Safety Cases submitted are for the most part prepared by the operator's personnel, although some use is made of consultants for specialist work such as consequence modelling, particularly by smaller companies. In assessing the Safety Case, the HSE is able to bring to bear the full range of expertise required. It possesses this expertise in-house. However, it does make use of consultants to assist with peak workloads.

17.14   The Safety Case is concerned with management and software as well as with hardware, as indicated by the information on management and management systems required in para 4 of the Schedule. Amplifying these requirements Dr Sefton stated:

"Information should be given which details operation and revision of safety policies; the setting or adoption of design and construction standards; quality assurance arrangements, operating procedures, training, management supervision, monitoring, staff welfare and management structure. All these separate elements are necessary to describe fully a system of management control and the report should give some indication of activity within every element. Control in the above spheres of activity requires:

(a) identification of work required to achieve the desired objectives; (b) the establishment of standards for described activities; (c) performance measurement to assess the degree of compliance with standards; (d) evaluation of performance over time which is communicated to accountable persons; and (e) the means to correct deficiencies in performance standards."

17.15   Dr Sefton was questioned on the way in which HSE goes about assessment of management and management systems. He replied that it was not possible to separate hardware and software concerns in the manner implied in the questioning. Scrutiny of preventive mechanisms might reveal weaknesses in management controls. Any perceived deficiencies in this area would be taken up by the local inspector. He stated:

"It is one of the skills of inspectors, to be able to interrogate the operations of companies to look not only at the hardware that is easy to see and easy to look at, but to get under the skin of a company, to ensure that they are setting the appropriate standards, that they do know what are the potential problems that they have, and that they are monitoring and assessing what they are and what they are doing. Much of the training of inspectors is associated with that. You cannot have one without the other. You cannot simply look at the hardware; you have to look closely at the management control of that hardware and of the hazards associated with it."

Asked whether an inspector would take the matter as high up the management tree as necessary, Dr Sefton replied:

"I think if we were convinced that there were failures of management we would go very quickly to the very highest level of management. I do not think inspectors would delay going to see and deal with the highest level of management. Once they had evidence of management failings it is no good talking to a foreman or works manager if the failing is a failing of the direction of the company. The great skill of inspectors, I suggest, is identifying quickly the failings that are leading to inadequacies on the ground and identifying where in the overall management structure the weakness is and homing in on it as quickly and effectively as possible."

If the local inspector required assistance on matters of management, he could call for specialist advice from the APAU.

17.16   The HSE witnesses clearly thought that the CIMAH Regulations had been a success. Mr Rimington called the regulations "a major step forward", but also said that when they were first brought in, he could not have said with confidence that they would produce the results which they had in fact produced. Dr Sefton too thought the regulations had largely achieved their aims. He believed this success owned much to the high level of technical expertise which the HSE had deployed and which the industry respected. It had shown an understanding of the issues of managing industry. Another HSE witness, Dr A F Ellis, Deputy Chief Inspector of the Technology Division, which includes the Major Hazards Assessment Unit (MHAU), was asked whether he thought the Safety Case was working out well, and in particular the role of quantification; he believed it was.

The Norwegian model

17.17   The Norwegian offshore regime, described by Mr Ognedal, has developed in the same general direction. The requirement for some form of risk evaluation is a long-standing one. The Regulations Concerning Safety Related to Production and Installation in 1976 contained a requirement that if the living quarters were to be located on a platform where drilling, production or processing of petroleum was taking place, a risk evaluation should be carried out. It was Mr Ognedal's recollection that at this date such an evaluation would have been mainly qualitative. In 1976 the NPD rejected a design for the Statfjord B platform as a copy of Statfjord A and required the living quarters to be put on a separate platform. In 1977 Mobil put forward a new concept of a platform integrated but with separation of the accommodation by a 6 hour rated firewall, which was accepted. The Statfjord B exercise influenced the legislation which followed.

17.18   The move to a more quantitative approach came with the Guidelines for Safety Evaluation of Platform Conceptual Design published in 1981. These centred around the provision of a shelter area, required the conduct of a concept safety evaluation (CSE), and specified numerical acceptance standards. A design accidental event was defined as one which did not violate any of the following 3 criteria:

"(a) at least one escape way from central positions which may be subjected to an accident, shall normally be intact for at least one hour during a design accidental event

(b) shelter areas shall be intact during a calculated accidental event until safe evacuation is possible

(c) depending on the platform type, function and location, when exposed to the design accidental event, the main support structure must maintain its load carrying capacity for a specified time." (clause 5.2).

The following categories of event were required to be evaluated, where relevant: blow-out; fire; explosion and similar incidents; falling objects; ship and helicopter collisions; earthquakes; other possible relevant types of accident; extreme weather conditions; and relevant combinations of these accidents. It was required that based on these design accidental events a set of design accidental effects should be specified, expressed in terms of heat flux and duration; impact pressure, impulse or energy; and acceleration. Explicit numerical acceptance criteria were stated:

"In practical terms, it may be considered necessary to exclude the most improbable accidental events from the analysis. However, the total probability of occurrence of each type of excluded situation (see 4.1.3) should not by best available estimate, exceed 10^{-4} per year for any of the main functions specified in 5.2, 5.5 and 5.6."

"This number is meant to indicate the magnitude to aim for, as detailed calculations of probabilities in many cases will be impossible due to lack of relevant data." (clause 4.2.2).

In effect, therefore, since there were some 9 categories, the requirement was that the frequency of the totality of events more serious than design accidental events, which are termed residual accidental events, should not exceed 9×10^{-4} per year, or in round figures 10^{-3} per year. These numerical criteria have been applied with flexibility.

17.19   The role of this FSA in the regime, as described by Mr Ognedal, is that the Guidelines are superior to the other, more prescriptive regulations. He confirmed that this meant that if a regulation laid down a particular requirement but risk analysis indicated that it was not necessary, an exemption from the regulation could be granted. Conversely, the analysis might show that the minimum requirement in the regulation was not sufficient.

17.20   The 1981 Guidelines were still in force but were to be replaced in 1990 by the Regulation on Risk Analysis, currently in draft. These new regulations require that safety analyses should be carried out through all phases from concept to operation, but the choice of the methods would be left to the operator. The new regulations would no longer contain a stated numerical acceptance criterion. Instead, the operator would be required to establish its criteria before the start of the conceptual design. Mr Ognedal stated that one of the reasons for the change was to "avoid further number game discussions". He affirmed that in making this change the NPD was not abandoning its original approach but building on it. The acceptance criteria required would not be less stringent. The whole philosophy underlying the legislation is one of progressive improvement.

17.21   The system operated by Statoil was described by Mr O J Tveit, a senior engineer with the company. In addition to carrying out a CSE at the conceptual stage, a total risk analysis (TRA) is performed at the detail design stage; this latter is an assessment developed by Statoil itself.

DEn practice and intentions

17.22   Prior to Piper there was no requirement in the British offshore regime for an FSA dealing with the whole range of major hazards. When I look at what the DEn had done, including the regulations and the associated guidance, to which certifying authorities work, and at the evidence given by Mr Petrie, there seems to me to be an imbalance between the attention given to the threats to the platform from environmental conditions and ship collision and those from the hydrocarbons. The approach which Mr Petrie described did not impress me as an effective one for the identification and control of the major hazards from the hydrocarbons at high pressure. He agreed that a large proportion of the inspection effort was in fact concerned with high pressure plant and that familiarity with pressure systems was a prime skill required in an inspector, yet his inspectorate was not strong in this area.

17.23   That this failure affected large parts of the UKCS was illustrated by the evidence of Mr A J Adams, a Principal Pipeline Inspector in the Safety Directorate of the PED, that prior to Piper there were some 70 risers, out of about 400 covered by the new ESV regulations, which did not have a true ESV and which required changes to the valve itself or to its actuator, control logic, etc, to make it such.

17.24   The DEn presented a discussion document on Formal Safety Assessments of Offshore Installations, which was spoken to by its author, Mr E J Gorse, a Principal Inspector and head of the section dealing with the auditing of the work of the certifying authorities. The document was in 2 parts, the first dealing with the principles of FSA and the second with factors to be taken into account, and was meant to cover both hardware and management aspects. The first part listed installations to be covered, hazards to be considered, techniques which might be used and project stages at which assessments should be carried out. It stated that there should be written procedures for undertaking FSA and that the outcome of the FSA should be documented and subject to independent regulatory review. The second part gave more detail of some of the techniques, including HAZOP and QRA. In effect, therefore, the document created a requirement for something analogous to the onshore Safety Case. However, the document was perceived to be weak on management and human factors aspects and Mr Gorse was questioned at length on this.

17.25   With regard to the regulatory review of such an FSA, it was the intention that the hardware, or technological, aspects, should be integrated into the certificate of fitness regime and that this aspect of the safety assessment should be taken into account when the certificate was issued. The assessment of the hardware aspects of the FSA would be done by the certifying authority. The importance of covering management as well as hardware was recognised, but the Department was still working on how this aspect should be assessed; it was "early days". An engineer was being seconded from the APAU on a permanent basis to assist. There was as yet no concluded view except that the assessment of written procedures and human aspects would be subject to some form of independent assessment. Questions were also asked on the expertise available within the DEn on FSA, QRA and management assessment. Mr Gorse stated that the Department did not possess the expertise to cover the whole range of FSA and that it had no expert on QRA or on management aspects.

17.26   The introduction to the discussion document described it as a major step forward, but despite the similarity between the FSA described and the onshore Safety Case, the document made no reference, even in the bibliography, to the CIMAH Regulations. Mr Gorse said that there were so many references which might have been quoted and it was necessary to be selective. This is in line with Mr Petrie's attitude to the CIMAH Regulations which I consider in Chapter 22.

17.27   It was made plain by Mr Petrie that in the regime which he envisaged an FSA would complement rather than replace regulations. As far as concerns the kind of regulation, he was in principle in favour of regulations of the goal-setting type.

However, although the Department had reviewed the body of regulations, it had not started on the task of formulating goal-setting regulations. No existing regulations had as yet been amended to goal-setting form. Whilst manpower constraints were a factor in this, another factor was the question of the balance between goal-setting and prescription. His position was that he was reluctant to lose the ability to make prescriptive regulations, though they would not be used unless there was a clear need for them. In any case he thought the difference was not clear-cut. It appeared that the Department continued to be attached to prescription of hardware. This was illustrated by the discussion document on Fire and Explosion Protection, considered in Chapter 19. This document was intended to fit in with a regime which was moving towards use of FSA. It still contained numerous prescriptive requirements for hardware, albeit some were expressed as default requirements. A similar approach underlay Mr Petrie's comments on guidance notes. He saw these as setting a minimum standard. It was put to him that there seemed to be an iron law that material intended as guidance came to be interpreted as mandatory. He agreed there was frequently a misconception about the status of guidance, but believed the present situation struck the right balance.

17.28   Mr Gorse confirmed that the FSA envisaged in the discussion document would apply to mobile as well as fixed installations, in fact to all installations, including floating production vessels and multi-purpose vessels.

Parties' submissions on an FSA

17.29   UKOOA submitted that the operator should be required to carry out an FSA, equivalent to a CIMAH Safety Case, in a planned manner at specific stages of the project such that the findings could be incorporated into the design or any proposed change in operating activity. The operator should define the design accidental events and the acceptance criteria. Quantitative methods should be used where appropriate. This FSA should be done by company personnel with the outside consultants confined to specialised work such as consequence analysis or QRA. Specifically, it was submitted that the following features should be dealt with in this manner:

> Management systems; need for a safe haven and its location, protection and facilities; location of accommodation and its protection against smoke; location and protection from smoke of control room, radio room, and emergency command post; number, location and protection of risers; subsea isolation valves; fire and gas detection systems; protection against fire and explosion; escape routes and embarkation points; evacuation and escape system.

17.30   Further, UKOOA proposed that the regime should move to one based entirely on the single regulation for an FSA and that other regulations would then be unnecessary. However, if this was not acceptable, the regime should at least cease to be based on prescriptive regulations and should move to one based on goal-setting regulations with compliance demonstrated by FSA.

17.31   The submissions of the Trade Union Group, the Piper Disaster Group and the Contractors' Interest all supported the concept of an FSA or Safety Case applicable to both new and existing installations, though they differed in the extent to which mandatory QRA should be required.

17.32   The Trade Union Group submitted that the Safety Case should be brought in forthwith by implementing the CIMAH Regulations offshore. This proposal was spoken to by Dr V C Marshall, a consultant, and its implications were examined with Dr Sefton.

An offshore Safety Case

17.33   I am convinced by the evidence that an FSA is an essential element in a modern safety regime for major hazard installations and that it has a crucial role to play in assuring safety offshore. Not only was there a consensus on this but also a large measure of agreement on how the matter might be taken forward. This consensus was confirmed by the parties' submissions. I consider that this FSA should take the form of a Safety Case.

17.34 The regime should have as its central feature demonstration of safe operation by the operator. To this end there should be a requirement for a Safety Case, based broadly on the CIMAH model for onshore installations. The CIMAH and Norwegian models show that this is both practical and desirable, the DEn was moving in this direction and it was in essence what UKOOA proposed.

Nature and purpose of Safety Case

17.35 Primarily the Safety Case is a matter of ensuring that every company produces an FSA to assure itself that its operations are safe and gains the benefits of the FSA already described. Only secondarily is it a matter of demonstrating this to the regulatory body. That said, such a demonstration both meets a legitimate expectation of the workforce and the public and provides a sound basis for regulatory control.

17.36 Both the evidence which I have already described and that which I will describe later make it clear that safety is crucially dependent on management and management systems. The Safety Case should show among other things that the company has a suitable safety management system. I defer further consideration of this aspect to Chapter 21.

17.37 The offshore Safety Case, like that onshore, should be a demonstration that the hazards of the installation have been identified and assessed, and are under control and that the exposure of personnel to these hazards has been minimised. I envisage that the general approach of the offshore Safety Case will be similar in many respects to that onshore. However, there will also be significant differences. In the offshore case the demonstration that the hazards are under control should include as a central feature a demonstration that the threat from these hazards to the arrangements for refuge for, and evacuation and escape of, personnel in the event of an emergency, is under control. The Norwegian regime follows this approach. I consider these matters further in Chapters 19 and 20.

17.38 An installation needs to be self-sufficient in providing protection for personnel. The Safety Case should demonstrate that it possesses a temporary safe refuge (TSR) and escape routes which will endure for a sufficient time to allow safe and full evacuation. I consider these matters further in Chapter 19. It is difficult to see how such a demonstration could be done other than by QRA and accordingly it is proposed that QRA be required. This requirement therefore goes beyond what is required onshore. It is clearly practical, since it is included in many onshore Safety Cases and is the basis of the Norwegian CSE. It is considered justified for offshore installations because large numbers of people not only work but live on them; the risks on the installations are relatively high; it is expected that the proportion of cases where the benefit of the QRA is marginal will be outweighed by those where it is substantial; the installations are much less heterogeneous than those onshore; they are substantial installations which justify the resources required to perform the QRA; and, not least, because in one tragic instance the hazards have been realised. It is proposed that the requirement should be for the estimation of the frequency with which there occur accidental events exceeding the design accidental events. In general, therefore, this is a requirement for explicit estimation of both frequency and consequences. However, it may be possible for certain hazards, perhaps even an appreciable proportion, to meet the requirement by a calculation of consequences only which makes it unnecessary to calculate the frequency. I consider QRA further in paras 17.48 *et seq.*

17.39 I have considered but rejected the proposal that a version of the CIMAH Regulations should be applied offshore with only those changes clearly essential for such application. It will be apparent that the offshore Safety Case which I envisage is sufficiently different from that onshore that this would not be the right way to proceed.

17.40   The Safety Case should normally be prepared primarily by company personnel. I accept the argument that a company which is competent to operate an offshore installation should be competent to produce the Safety Case. Moreover, involvement of the company's own personnel is the best way to obtain the full benefits within the company and for the purpose of dialogue with the regulators. Similarly, it is desirable that the operator should deal itself with the QRA aspects of the Safety Case rather than contract them out. Familiarity with the system is essential for good QRA and companies often prefer to employ engineers familiar with the system and train them in QRA techniques rather than to call in risk analysts and acquaint them with the system. Moreover, use of company personnel allows expertise to be built up in-house. On the other hand consultants have a role in bringing an independent perspective and assisting with novel and specialist techniques.

17.41   The Safety Case should apply to both fixed and mobile installations. The question of the application of an FSA to mobiles was explored and no impediment was identified. It was the intention of the DEn to make the FSA which they proposed applicable to mobile as well as fixed installations.

Safety case for new installations

17.42   Onshore there is a requirement for a Safety Case both for new and existing installations. I believe that the same should apply offshore. There is little dispute about the benefits to be gained from a Safety Case for a new installation. For such an installation there is clearly great value in some form of CSE. The initial form of the Safety Case should have this character. As the design develops so should the Safety Case, taking on more the aspect of a TRA. It is intended that in the final form in which it is submitted the Safety Case should be based on detail design information. I note that the CIMAH Regulations require the onshore case for a new installation to be submitted not less than 3 months before the commencement of the activity (Reg 7(1)), which indicates that the case will contain detail design information.

17.43   It will be for the regulatory body to specify the precise stage in the project for submission of the Safety Case. It is clearly desirable that some preliminary assessment of matters related to the Safety Case be submitted early in the project, preferably on application for Annex B consent. The regulatory body should consult with the industry on this.

Safety Case for existing installations

17.44   I consider that a Safety Case should also be required for existing installations. This is the requirement onshore. The risks offshore are clearly no less. It is not acceptable that installations should be operated without a thorough assessment of what those risks are. While certain options are foreclosed once an installation is built, there will generally be a variety of measures, both hardware and software, which can be taken to improve safety if the risks justify them. Since in this case the full detail design information is available, the Safety Case will have the character of a TRA.

17.45   Safety Cases for existing installations should be brought in as rapidly as practicable, on a schedule to be determined by the regulatory body.

The continuing Safety Case

17.46   The Safety Case should be seen not as a one-off exercise but as part of a continuing dialogue between the operator and the regulatory body. I have already described the increasingly central role assumed by the Safety Case in the onshore regime for major hazards and envisage a similar role for the offshore Safety Case. It follows that the Safety Case needs to be kept up-to-date. It should be updated at regular intervals or if there is any material change affecting it. The most fundamental change will be a change of operator; an updating of the Safety Case is essential in this case. An updating should also be triggered if there is a major emergency on the installation, with or without precautionary evacuation; if there are major modifications; or if there is some major technological innovation or the discovery or improved understanding of a major hazard which might justify it.

17.47   Given that the Safety Case should be updated if there is a major modification to the installation, there will be a need for the regulatory body to define what constitutes a major modification for this purpose.

Quantitative risk assessment

Role and status of QRA

17.48   Accounts of QRA were given to the Inquiry by Dr Cox and Dr Hogh. I deal here with just one or two points in order to make clear the role which I envisage for it as an aspect of an FSA.

17.49   I endorse the emphasis placed by both witnesses on the fact that QRA is only one input to the decision-making process, though an important one. Its strength is that it provides a structured, objective and quantitative approach. It gives a better understanding of the hazards and of the measures needed to control them. The operator is required by the HSWA to take all reasonably practicable measures to ensure safety. QRA is a prime means for the operator to demonstrate firstly to itself and secondly to the regulator that it has done this and thus provides a good basis for the dialogue between operator and regulator. It should not be used, however, in isolation or as an automatic mechanism for decision-making. The point is made in one of the documents on QRA published by the HSE, "Quantified Risk Assessment: Its Input to Decision-making", quoted by Dr Cox:

"QRA is an element that cannot be ignored in decision-making about risk since it is the only discipline capable, however imperfectly, of enabling a number to be applied and comparisons of a sort to be made, other than of a purely qualitative kind. This said, the numerical element must be viewed with great caution and treated as only one parameter in an essentially judgmental exercise." (para 10).

17.50   I am aware that QRA has been a matter of some controversy. There was general agreement that it is a complex subject. However, as Dr Hogh said, complexity is not synonymous with difficulty. Whatever may have been true some 10 years ago, both Dr Cox and Dr Hogh considered that there was now no serious problem in obtaining the data required to estimate frequency or models to estimate consequences; the area of human factors was acknowledged to be one where improved techniques were desirable. Dr Hogh in fact described QRA as a normal tool of project management. In giving this evidence both witnesses were referring to the application of QRA offshore as well as onshore. I am satisfied that there is no impediment to the use of QRA offshore. I agree, however, that it is desirable to be quite open about the uncertainties inherent in QRA and to take these into account in its conduct and evaluation, using the methods of sensitivity analysis described by the witnesses.

Regulatory uses of QRA

17.51   HSE's view of the role of QRA in the regulatory regime was put by Dr Ellis and Dr R P Pape, Head of the Major Hazards Assessment Unit. For nuclear installations QRA is a normal part of the Safety Case. It does not have this status for process plants.

17.52   HSE accepts that there is some controversy about the use of QRA and has recently published 3 documents to make its views known and to stimulate discussion. Dr Ellis quoted from the same publication as Dr Cox:

"It is therefore important to be able to predict what could happen and as far as possible how likely or unlikely it is - as well as recording what has actually happened, and then to see how best to control, and if possible reduce, the risks that are identified. For this QRA is an indispensable element, but one to be used with caution and not applied mechanistically to demonstrate compliance with legislative requirements." (para 15).

17.53 HSE's own interest in QRA arises because it is an organisation which regards it as important to found any legal or political judgement as firmly as possible on a rigorous scrutiny of the facts, using the available techniques. It is conscious that it is dealing with technologically based industries or scientifically numerate organisations which expect a structured and logical approach. Equally, it is conscious that not all health and safety problems can be reduced to mathematical terms. Nevertheless, it believes that quantification, or in some cases just the attempt to quantify, imposes a discipline beneficial to safety. Dr Ellis drew a distinction between quantification and full QRA. He agreed there was enthusiasm in the HSE for the former, while for the latter there was "cautious enthusiasm". In some cases the full process of QRA is not necessary. The quantification of the potential consequences of an accident may be sufficient. HSE's views were much stronger on quantification of consequences than on full QRA. He quoted as an example of HSE's attitude to quantification the following extract from the guidance notes to the CIMAH Regulations:

"Whilst it may be possible for manufacturers to write a safety case in qualitative terms, HSE may well find it easier to accept conclusions which are supported by quantified arguments. A quantitative assessment is also a convenient way of limiting the scope of a safety case by demonstrating either that an adverse event has a very remote probability of occurring or that a particular consequence is relatively minor." (para 112).

Dr Ellis stated that while QRA might not be specifically required by regulations under the HSWA, the general requirements of that Act could imply a need for QRA where it is likely to be worthwhile. He agreed that in order to decide whether an installation was acceptably safe, it is reasonable to want to know the level of risk which it poses. Asked whether the HSE had the powers to require a QRA from an operator, Dr Ellis said the question was difficult to answer; it had never tried to enforce such a requirement.

17.54 The selective use of QRA by regulatory bodies was supported by Dr Hogh as providing a framework for dialogue. However, the industry had been resistant to the blanket application of QRA to existing onshore major hazard installations as a requirement of the CIMAH Safety Case. It was an essentially futile exercise unless carried out for a defined purpose.

Acceptance standards for QRA

17.55 The practice of QRA requires acceptance standards. There is more than one form of acceptance standard. Examples are accommodation endurance times, equipment availability targets and risk criteria. As far as risk criteria are concerned, I would expect the general approach to be that described in the HSE discussion document on the tolerability of risk and shown in Fig 17.1, which was introduced by Dr Hogh and endorsed by the other witnesses. The upper line is that above which risk is intolerable and action must be taken, the lower region is that in which risk is negligible and no action is required, while in the intermediate region the requirement is to reduce the risk "as low as reasonably practicable" (ALARP). This latter implies a cost-benefit analysis. In formulating risk criteria, due regard should be had to risk aversion, the aversion which society has to major accidents. Risk aversion should receive recognition not only in setting the upper bound of what is acceptable, but in the cost-benefit analysis.

17.56 It is normal practice that acceptance standards for QRA are set by the operator. This accords with the fact that QRA is generally an activity undertaken voluntarily

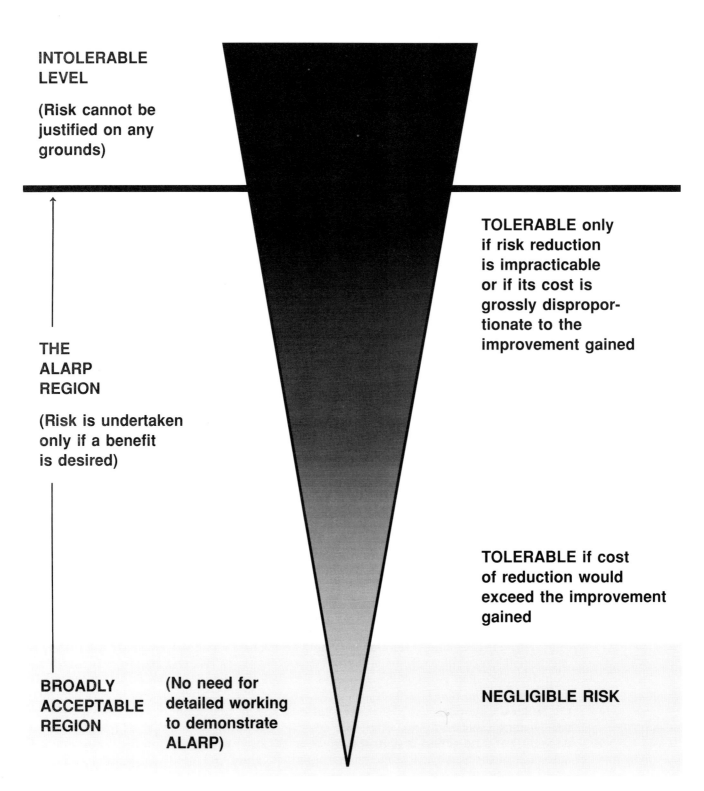

INTOLERABLE LEVEL

(Risk cannot be justified on any grounds)

THE ALARP REGION

(Risk is undertaken only if a benefit is desired)

BROADLY ACCEPTABLE REGION

(No need for detailed working to demonstrate ALARP)

TOLERABLE only if risk reduction is impracticable or if its cost is grossly dispropor- tionate to the improvement gained

TOLERABLE if cost of reduction would exceed the improvement gained

NEGLIGIBLE RISK

Fig. 17.1 Levels of risk and the ALARP principle.

to demonstrate compliance. The HSE has published documents on risks and risk criteria but as guidance. I consider that this is the approach which should also be adopted offshore.

17.57   I propose, however, one exception to this general principle. The Safety Case involves a demonstration that the frequency of events which threaten the endurance of the accommodation, or TSR, will not exceed a certain value. In order to provide at least one fixed point in the regime, both the minimum endurance and the frequency with which there is a failure of such endurance should be specified by the regulatory body, at least in the first instance. This proposal is described further in Chapter 19.

17.58   I fully endorse the view expressed that acceptance standards, including risk criteria, should be interpreted with flexibility by the regulatory body.

17.59   This is not to say, however, that the acceptance standards should not be tough; they should be. In the regime proposed these standards will be one of the main pressures for improvement. Unless they are set sufficiently high, they will not be effective. It is my intention that the regulatory body should require acceptance standards which will result in real improvements in safety. In particular, there needs to be a reduction of the risks from major accidents. The Inquiry did not go into risk comparisons, but it is clear that the historical risk to the workforce in the UKCS is now dominated by a single accident, the Piper disaster. A similar situation pertains in the NCS following the Alexander Kielland disaster in which 123 died.

17.60   Whilst in general QRA requires some standard of comparison, it does not always involve absolute risk criteria. The point was made that there is a distinction between inherent features such as layout and add-on features such as a protective system. It was not possible not to have a layout, the question is to choose between different layouts, and QRA may be used to assist the choice by comparing safety aspects, without necessarily using absolute risk criteria. On the other hand a protective system is in a sense an optional extra and in this case the use of QRA to aid this decision implies the use of some absolute criterion.

Application of QRA to existing installations

17.61   As I have already stated, I propose both that the Safety Case should involve QRA and that there should be a Safety Case for existing as well as for new installations. It therefore follows that I am proposing QRA for existing installations. I have already mentioned Dr Hogh's comment that the industry had been resistant to the blanket application of QRA to existing onshore major hazard installations as a requirement of the CIMAH Safety Case. I am satisfied that the QRA in the Safety Case which I propose has a well-defined purpose. In brief, it is to assess the risks, to identify and assess potential safety improvements, and to ensure that the TSR meets the standard set.

Safety assessment and regulations

17.62   I now turn to consider some other aspects of FSA in the regime. So far I have deliberately confined myself to the question of an FSA, or Safety Case. I now consider FSA in its more general sense, and in particular as an activity which may be undertaken to demonstrate compliance with legislation.

17.63   The regime should not rely solely on the Safety Case. I reject the argument of UKOOA that the only regulation should be one requiring an FSA as going much too far. In general, any large system or problem is usually best handled by breaking it down into more manageable parts, in some form of hierarchy. I propose that the regulation requiring the Safety Case should be complemented by other regulations dealing with specific features. This is in accordance with the approach taken onshore, where the regulations continue to exist alongside the Safety Case, and this not just for historical reasons but as a matter of policy. With regard to the type of regulation, the Safety Case would sit well with regulations which set goals rather than prescribe solutions. These regulations would complement the Safety Case by setting intermediate goals and would give the regime a solidity which it might otherwise lack.

17.64   Construction of the installation, fire and explosion protection, and evacuation, escape and rescue are all areas where I consider it appropriate to retain regulations, though in goal-setting form. One method of demonstrating compliance would then be by FSA.

17.65   Since it may appear that these are all areas which might be covered by the Safety Case, it may be helpful to give an example of a specific requirement of a goal-setting regulation. It is proposed that the regulation dealing with fire and explosion protection should contain a requirement for a reliability assessment of the fire pumps. The precise means by which fire pump availability is to be achieved would be left to the operator, but it should be able to demonstrate independently of the Safety Case that, at least for all eventualities other than disablement by the accident itself, availability targets have been specified and will be met. While arguably this, like almost everything else, could be left to the Safety Case, it is inconceivable that there should not be such an assessment, and it is therefore entirely appropriate to cover it by means of a regulation. What the Safety Case contributes is a set of major hazard accident scenarios against which the design of the system can be further assessed.

17.66   It is envisaged that the operator will demonstrate compliance with a goal-setting regulation by a variety of means. It may do so by reference to guidance, or to in-house standards, or to FSA or to some combination of these.

17.67   The transition to the new regime cannot take place overnight. I propose that there should be a regulation requiring a Safety Case and that this should be complemented by a limited number of further, defined regulations, but beyond this it must be for the regulatory body to develop the regime in accordance with the principles outlined. As regards existing regulations and guidance during the transition, I do not envisage any wholesale revocation of regulations or withdrawal of guidance, but suggest that the regulatory body advise the industry of those regulations to which it is prepared to grant exemption in the light of a demonstration of a satisfactory alternative in the Safety Case.

Safety assessment in the regime

17.68   The operation of the regime would then involve FSA

   (i)   in compliance with the regulation for a Safety Case;

   (ii)   in compliance with any other regulation requiring a safety assessment;

   (iii) as a means of demonstrating compliance with a goal-setting regulation; and

   (iv)   as a means of demonstrating compliance with the HSWA.

In the first two cases the safety assessment would be mandatory, in the last two it would be voluntary.

17.69   In some cases a goal-setting regulation will contain a requirement that the design should be subject to an analysis to demonstrate that it is satisfactory. I describe below my proposals that there should be analyses of fire risk and of evacuation, escape and rescue and I have already proposed that there should be a Safety Case. I envisage that the Safety Case should test the design in respect of major hazard accident scenarios and that the analysis should test it at least in respect of all matters short of those scenarios. I have given above (para 17.65) the example of fire pump availability as part of a fire risk analysis. Similarly, an analysis of evacuation, escape and rescue would test among other things the arrangements for man overboard (MOB) incidents.

Safety assessment and the regulatory body

17.70   My proposal to assign to FSA in general and to the Safety Case in particular a central role in the regime has obvious implications for the regulatory body. Although this was considered in the Inquiry largely in relation to the ability of that body to evaluate an FSA, the question is much wider than that. Onshore the Safety Case has come to form the basis for the HSE's inspection activity on major hazard installations. Similarly, offshore the regime which I envisage is one in which the emphasis moves to audit of the operator's systems and in which the Safety Case provides a starting point for such audit. The regulatory body must be one which is not only able to evaluate the Safety Case itself but is at ease with this whole approach.

17.71   However, considering the narrower issue of the ability of the regulatory body to evaluate the operator's Safety Case, it is clear to me that this must be done by a single regulatory body. A strong plea for a single point of contact was made by Mr Ferrow, who argued that the FSA is an integrated whole, comprising hardware and software aspects which interact, and it would not be satisfactory to split the evaluation between different bodies. The weight of the evidence pointed to the need for a single body competent in FSA, confident in its own ability and capable of being flexible, and credible with the industry. The nature of the decisions involved in responding to FSA is such that they cannot readily be delegated and the attempt to do so is liable to result in divided responsibility, excessive caution and undue delays, whereas what is wanted is an authoritative and prompt response. Separation of hardware and software is artificial. It follows that I reject the approach which the DEn was intending to adopt.

Chapter 18

The Prevention of Incidents Causing Fires and Explosions

Introduction

18.1   In order to meet its duty to provide, as far as reasonably practicable, a safe system and place of work an operator has to seek to prevent incidents which can lead to fires and explosions, to mitigate the consequences of incidents which do take place, and to provide a safe method of evacuation and escape and rescue. In this chapter I will examine the first of these, the prevention of incidents.

18.2   The nature of the process fluids in any hydrocarbon plant is such that they will burn readily and, when combined with air, can lead to a mixture which will explode if ignited. Consequently the first objective of any safety policy and programme is to prevent incidents which may cause a loss of containment of hydrocarbons. This is particularly important in the offshore industry as mitigation of the effects of fire and explosions and evacuation or escape of personnel are inhibited by the remoteness and isolation of oil platforms. I will consider the specific lessons which can be drawn from the disaster on Piper in relation to preventing such incidents, with particular reference to the PTW system, the control of the process, the introduction of modifications to a platform process, and the investigation of accidents.

18.3   However important those lessons might be no one incident, even one as disastrous as that on Piper, can point up more than a few important improvements in offshore safety. Equally in practice exactly the same accident hardly ever repeats itself, so management needs to address the spectrum of possibilities and not just seek to prevent recurrences. Accordingly I have in the later part of this chapter examined how offshore operators approach the issue of managing in order to prevent incidents which can lead to emergency situations.

Permit to work systems

18.4   I have set out in earlier chapters that the PTW procedure on Piper failed as a component of a safe system of work. It suffered from deficiencies in regard to the actions taken to suspend a permit, the absence of a procedure for locking off isolation valves, the lack of cross-referencing, the lack of a procedure for handover of permits at shift change, inadequate training of contractors' staff and an ineffective auditing system. Evidence on those aspects of PTW systems was presented by Mr S R Kyle, Environment and Safety Co-ordinator for the Brae Operations of Marathon Oil UK and chairman of UKOOA's working group on permits to work, supported by examples from the procedures of particular operators. Mr G H Davies of the HSE set out the practices in onshore industry, and Mr T J Scanlon, who had been employed by Wood Group Engineering, a major contracting company, gave evidence on behalf of the Contractors' Interests.

Suspending a permit

18.5   The bulk of maintenance work offshore is carried out on day-shifts only, normally extending from 06.00 hours to 18.00 hours. Inevitably there are many tasks that cannot be completed in one day. Additionally there are occasions where work will be interrupted for longer periods, mainly when the platform has to await the supply of spare parts from the shore. It is essential that any PTW system incorporates a procedure to ensure that in such circumstances the equipment being worked on is retained in a safe condition and no attempt is made to use it in operations.

18.6 Mr Kyle explained that the key procedural steps to achieve this are

— at the end of the shift when the work is to be suspended, the Performing Authority should, after inspecting the site, sign the permit to the effect that the work is suspended and return it to the Designated Authority;

— the Designated Authority, or one of his delegated operators, should inspect the site to check that conditions are such that the work may be safely suspended and should then sign the permit accordingly;

— the permit, together with other suspended permits, should be located in a prominent position in the control room or permit office clearly labelled as "suspended";

— to ensure that equipment on which work is suspended is not used the PTW system must be supported by a secure method of isolation (see para 18.10);

— prior to re-issuing the permit the Designated Authority must ensure that the site is inspected again to make certain all required isolations are still in place.

18.7 Mr Davies explained that suspension of PTWs is not a practice widely recognised in the onshore oil and chemical industries and the common practice is for the permit to be properly cancelled and re-issued when work is to resume. Whichever approach is adopted the principle that should apply is that, when work ceases for a temporary period, the equipment must be left in a safe state. While it is not being worked on all the specified precautions, such as isolations, must remain in place and when work is to re-start those precautions must be checked. The importance of this principle was emphasised by a recent survey in the onshore chemical industry which found that in 25% of the PTW systems examined there were inadequate handback procedures.

18.8 I am satisfied that the procedural steps outlined by Mr Kyle are consistent with the principle set out by Mr Davies of the HSE. They are simple to operate in practice and straightforward to incorporate in any PTW system. They should be used in all offshore PTW systems.

Locking off isolation valves

18.9 It was Mr Kyle's evidence that it is essential that a PTW system is supported by a secure method of isolation and that necessary security should incorporate the locking off of isolation valves such that they cannot be accidentally or inadvertently opened. One such method used by Marathon Oil UK and demonstrated to the Inquiry involves a system of locks and key safes, whereby all the keys held by both Designated Authority and Performing Authority have to be returned to the key safe box before the box can be opened and de-isolation effected. Several offshore operators have similar effective methods of securing isolation. Mr Davies was aware that locking off systems are frequently employed in the onshore industry.

18.10 I consider that a physical locking off system should be an integral part of any PTW procedure because of the security it provides against inadvertent or unauthorised de-isolation. Certainly, if such a system had been employed on the Piper platform, it would have prevented the operating staff from opening the isolation valves which admitted condensate to the A condensate pump, which led to the leak of hydrocarbons.

Cross-referencing of permits

18.11 Both Mr Kyle and Mr Davies were agreed that where jobs involving separate teams of people interact, particularly in relation to the isolation of equipment, the permits for those jobs must be cross-referenced one to the other to ensure that no interaction takes place which might threaten the safety of the personnel or the platform. The responsibility for recognising the potential for interaction rests with the Designated Authority who must be supported by a good communication system on a daily basis to ensure that planned critical activities are made known to all affected personnel.

18.12   Where it is necessary for more than one task to be carried out on one piece of equipment or system utilising the same isolations a means of ensuring the integrity of the isolations until all jobs are complete must be established. In addition to cross-referencing the permit forms, a physical means of achieving this should be employed. Various methods are known and available to the industry, such as multiple locks or keys and key safes used in conjunction with the physical locking off methods I have recommended in para 18.10. At the completion of any one task the locking mechanism specific to its permit is removed but de-isolation cannot be effected until all such mechanisms for every permit have been removed.

Handover of permits

18.13   Mr Kyle considered that company procedures should ensure that adequate arrangements are in place for handover between Designated Authorities at shift change, that those should include specifically effective means of communicating the status of all active and suspended permits, and that sufficient time is available to achieve an effective handover. The means of communicating the status of permits may be by permit log books, permit files, or display boards. It would be expected that both Designated Authorities, outgoing and incoming, should review each permit together. Mr Davies' advice was that the handover arrangements should include also that the incoming lead operator sign for the continuation of any permit and no handover should rely on memory alone or be solely verbal.

Training in the PTW procedures

18.14   It was Mr Kyle's experience that PTW systems are only as good as the care and competence of the people who operate them. It is therefore essential that all persons who are required to operate the procedures and the tradesmen who work under permits are adequately trained. Specifically, detailed and formalised training in the PTW system for the platform on which they are to work should be given to both Designated Authorities and Performing Authorities, and they should be formally assessed prior to their appointment. Records of all PTW training should be maintained. Designated Authorities require in-depth training covering all aspects of the PTW system and procedures. Such training would take 2-3 days full time, concluding with a formal written examination. However, examination alone will not guarantee competence and it is important that individuals have demonstrated adequate experience, local knowledge and their ability to discharge their responsibilities competently prior to appointment by the OIM. It was expected that formal training of a similar nature for a Performing Authority would take one day full time. In this case also the appointment should be confirmed by the OIM.

18.15   Mr Davies' view was that formal training in the operation of the PTW system was a necessity and the system was unlikely to succeed in providing a safe system of work if it relied entirely on on-the-job training by another, however experienced, operator and if there was no formal assessment of how the training had been absorbed.

18.16   Mr Davies explained that onshore particular attention has to be paid to the training of contractors' personnel who may not be familiar with the plant's hazards, with the work procedures or to any extent with the detailed equipment. That training should be provided by the occupier of the installation - the operator in offshore terms - and the training for contractors should be specific to the installation on which they will work. This view was strongly supported by Mr Scanlon who argued that permits and all they entail should be explained by the operators as "contractors are not the best people to do that". Mr Kyle's own company, Marathon Oil, require the same training and assessment of contractors' personnel who are going to act as Performing Authorities as it does of its own employees and that each Performing Authority undergo refresher training every 2 years. Marathon do not envisage contracting companies giving any training to their employees in the Marathon PTW system. Surprisingly Mr Kyle's evidence was that this subject, the training of contractors' personnel, had not been discussed in the UKOOA working group on PTW systems.

It had to take that overall culture into account in its design for that particular location. The PTW system was likely to fall into disrepute more easily if it was at odds with the overall systems of the installation. While most steps in a PTW system were common and there could be a standardisation of principles, there would need to be flexibility to allow individual variations. For example, the needs of large companies and small companies might differ. Any system must take account of the underlying philosophy of the company, how it controlled all its systems of work, whether they were maintenance procedures or operational procedures. While there were advantages in standardisation there were also advantages in having the system as job specific as possible.

18.24   The Trade Union Group and the Contractors' Interests as advocates for standardisation considered that the UKOOA objection on the grounds of differing management structures might be overcome by seeking a common organisational theme. The problem of large scale re-training was there in any case as PTW systems were being continuously modified to effect improvements. The UKOOA proposals on common procedures implicitly accepted that a substantial degree of standardisation was possible.

Intentions of the Department of Energy

18.25   The requirement for operators of offshore installations to have in place a PTW system is covered in the Operational Safety, Health and Welfare Regulations. In the light of its own investigation into the Piper disaster the DEn issued Safety Notice 16/88 in November 1988 asking operators to review the scope, operation and control of PTW procedures. This was followed in July 1989 by the DEn setting out for comment its intention to strengthen the statutory requirements for work permit procedures. Its proposals are for an extension of those areas of work on an offshore installation which must be covered by a work permit system and the clarification of administrative procedures which should apply. The latter have much in common with some of the detailed principles set out by Mr Kyle as to be included in the UKOOA guide.

Conclusions on PTW systems

18.26   It is clear to me that much needed to be done to improve the general standard of PTW procedures in the UKCS, demonstrated by the number of changes that have already been introduced by various operators, as the evidence to the Inquiry showed. Many operators formed specific teams to ensure that the lessons of the Piper disaster were analysed in relation to their own activities. That was a praiseworthy response by the industry. The deficiencies revealed in the Occidental PTW system were not new problems, either offshore or onshore, and the better PTW systems have avoided them by straightforward procedural steps.

18.27   I am not persuaded that the introduction of a standardised PTW system offshore is either necessary or desirable. The concept originates in seeking a solution to the problem of contractors' supervisors having to act as Performing Authorities. I am satisfied this problem can be safely overcome by full and adequate training but it is clear that the responsibility for and the cost of that training should rest with the operator of the installation and cannot and should not be undertaken by the management of the contracting company. While it is preferable that an operator's own employees act as Performing Authorities it is inherent in the nature of the way maintenance work is executed offshore that there will be the need for contractors' staff to act in that capacity.

18.28   Standardisation is not desirable because the PTW system must marry with the individual operator's safety philosophy, organisation and methods of doing work. Additionally the need to change PTW procedures in the light of audit findings would be inhibited and the implementation of improvements made cumbersome if the whole industry were subject to each detailed change. Although standardisation in detail is undesirable there is no doubt that there must be common principles underlying any PTW system. Accordingly I welcome the UKOOA proposals on guidelines applicable throughout the offshore industry and I look to that organisation to accept the responsibility for ensuring that those common principles are implemented by each and every operator.

18.29   Such an approach would be consistent with seeking to have goal-setting rather than prescriptive regulations. However two improvements to PTW practices are of such overriding importance they should be incorporated in the current review of the Operational Safety, Health and Welfare Regulations planned by the DEn. They are:

— all permit to work systems should incorporate a mechanical isolation procedure which involves, wherever practicable, the physical locking off and tagging of isolation valves to prevent their accidental or unauthorised opening;

— operators should be responsible for and undertake the training of all staff, the operators' own or those of contractors, in the detailed PTW procedures where those staff are required to act as Designated or Performing Authorities. The training should be recorded and staff should carry documentary proof of having undergone such training.

Control of the process

Piper Alpha

18.30   It was evident in examining the circumstances of the disaster that the Piper control system had some limitations when it came to handling a developing emergency situation. In reacting to the events in the condensate system nothing could be done from the Control Room. The lead process operator, Mr Vernon, had to leave the Control Room to take command at the site of the condensate pumps. In consequence the supervisor with the responsibility and authority to decide whether the developing situation warranted partial or complete shutdown of the platform was absent from the control centre. He was not aware of the gas alarms and other signals that preceded the explosion. The Control Room was more a monitoring and message station than a place from which the process could be controlled. While instrumentation in the Control Room showed the status of equipment and alarms, the actual panels from which equipment operation could be adjusted were located within individual plant modules. These panels gave more detailed information on the condition of the plant; and the equipment controls could be operated from them.

Modern control systems

18.31   Mr M Ashworth, a Senior Control Engineer of BP International, explained that the type of control system on Piper was out-dated. It was expensive to provide in terms of space, weight and cost. Equally it was expensive to operate and maintain, being manpower-intensive and based on older technologies. There had been a progressive development of control room facilities over the past 15 years, the overriding influence being the technological advancement in computers and data communication. The modern concept was to provide a single central control room. The instrumentation was designed to control the platform process within a defined safe operating range. By means of alarms operators were warned when this range was likely to be exceeded; and such warning was given in time to allow the operator to initiate corrective action. The Control Room gave a detailed display of all equipment and process conditions and contained the necessary means of control for both normal and emergency operation. This allowed the installation to be operated safely with greater efficiency at reduced cost. The control room should be manned at all times by an experienced and trained control room operator.

Emergency shutdown systems

18.32   Mr Ashworth also pointed out that, in a modern control system, the first objective of a properly designed ESD system was to prevent an uncontrolled or hazardous situation occurring, in addition to its wider known use in reducing the consequences of a hazardous event when activated manually by an operator observing an emergency situation. By continually monitoring the process equipment with a range of protective instrumentation and taking automatic action to shut down when a predetermined value is reached, the ESD system became the principal mechanism for setting the installation to a known and safe state by stopping the process and isolating electrical equipment. The ESD system was therefore one component in a network of equipment and procedures which were designed to prevent process incidents developing in such a way as to lead to a release of hydrocarbons, and to maintain safe operation in both normal and emergency situations. The philosophy was to control the process within normal bounds and to detect and restrain abnormal events before they could escalate into an emergency.

Regulation of control systems

18.33   Apart from the requirement to have an ESD system there is no regulation specifying the type of process control system that should be installed on an installation. Mr M Ognedal, Director of the Safety and Working Environment Division of NPD, gave evidence that in Norway also there was no regulation as to the control of the process. The operating company was responsible for control of the plant. It was in its own interests to avoid unnecessary use of the ESD system and the stopping of production.

Training of control room operators

18.34   Mr Heiberg-Andersen, the OIM of Statoil's Gullfaks C platform, explained that it was necessary to give control room operators thorough training before they took up their positions on a platform. He described the onshore training of process operators using a process simulator which was a full scale replica of the platform control room together with an operating computer. The Gullfaks Control Room operators receive training on such a simulator before being appointed to the platform. That training included specific training in responding to an emergency.

Conclusions on process control

18.35   I am satisfied that there is no need to specify by regulation the type of control system that should be used on North Sea platforms, as it is in an operator's own interests on safety and cost grounds to adopt the available modern technology. What is clear from the evidence presented is that the control room should be manned at all times and be in the charge of a person trained and qualified to undertake the work of a control room operator. The training of control room operators must include instruction, in a properly developed onshore course, in the handling of emergencies, and that should involve practice in simulated emergencies.

18.36   It is evident that there are a number of platforms in the UKCS which have control systems which are of the same age as that on Piper and therefore are less able to deal with an emergency than a modern control system. This is a matter which could not be tackled effectively within the Inquiry although it was clear that complete replacement of those systems is not practicable. As a minimum I recommend that alterations should be made such that key process variables, as established by the FSA, are capable of being monitored and controlled from the Control Room.

The installation of modifications

18.37   I have pointed out in Chapter 6 that the institution of the temporary process system to inject methanol into the Piper process to prevent hydrate choking left much to be desired. The deficiencies in that task could have led to a potentially serious process upset and the release of hydrocarbons.

18.38 The requirement for the safe management of the installation of modifications is well known throughout the oil industry, in part as a result of the investigation into the major onshore disaster at Flixborough in 1974, which was traced to the inadequate management of a modification to the process and equipment. However, the circumstances surrounding the methanol injection system on Piper demonstrate that offshore operating companies would do well to monitor continuously their systems for the design and installation of modifications.

Incident reporting

Fatal accident on Piper

18.39 In Chapter 14 I commented adversely on Occidental's investigation of the fatal accident to Mr Sutherland and on the dissemination of lessons from that accident. The investigation lacked thoroughness in that it did not seek out the root causes of the accident or try to determine whether the particular failures associated with the incident were endemic in the way operations were carried out on Piper. The lessons from the investigation were not properly communicated throughout the organisation.

Offshore industry practice

18.40 The Inquiry heard evidence on the practice as to the investigation of incidents.

18.41 In Conoco (UK) Ltd it is part of the safety policy that all accidents or near misses are automatically investigated. A "near miss" is defined as a near accident that could have involved serious injury or had the potential for serious damage to property or the environment. The chief executive has a system to ensure that he is informed immediately of every significant incident and virtually all accidents no matter how minor. He is kept fully briefed on the progress and results of any investigation. Any personal injury greater than "first aid" severity is discussed in fortnightly meetings of the company directors. I have to point out that the frequency of accidents in Conoco is so low that the commitment of time by the directors to the discussion of injuries is not burdensome. It is also company policy to disseminate incident reports up and down throughout the organisation.

18.42 It is also the policy of the Amoco (UK) Exploration Company that all accidents and dangerous occurrences are the subject of an investigation and report describing the incident, the cause and the proposed corrective action. That action is checked out by both line management and safety specialists to ensure it is adequate. Significant dangerous occurrences or accidents involving injury to either company or contractors' employees are investigated by a committee of non-involved staff, normally comprising a safety specialist, an operating supervisor and an engineer. Written procedures exist for the investigating team to follow and their report is distributed throughout the company and specifically to the chief executive officer. Once agreed, corrective actions are followed through until all are complete.

Conclusions on incident investigation and reporting

18.43 I am convinced that learning from accidents and incidents is an important way of improving safety performance. That view is commonly held throughout the UK offshore industry. In relation to preventing incidents which cause hydrocarbon leaks that could lead to fires and explosions I consider it would be useful if there was a systematic means by which what could be learnt from such accidents and near misses was shared by all operators. The regulatory body should be responsible for maintaining a database with regard to hydrocarbon leaks, spills and ignitions in the industry and for the benefit of the industry. The regulatory body should

    — discuss and agree with the industry the method of collection and use of the data;

    — regularly assess the data to determine the existence of any trends and report them to the industry;

— provide operating companies with a means of obtaining access to the data, particularly for the purpose of carrying out quantified risk assessment.

Managing to prevent incidents

Introduction

18.44 While I am convinced that the lessons from the investigation of the Piper disaster which I have set out above will lead to improvements in safety offshore I recognise that those improvements are inevitably limited in their scope by the circumstances of the disaster. The prevention of incidents in the industry at large is dependent on the approach and quality of the management of safety by each and every offshore operator. Accordingly I sought evidence in the Inquiry as to good industry practice in the management of safety. Mr R E McKee, Chairman and Managing Director of Conoco (UK) Ltd and Mr R A Sheppard, Vice-President of Production and a Director of Amoco (UK) Exploration Company Ltd gave evidence. As would be expected the approach of the 2 companies was different in detail and it would not serve any useful purpose to list those differences in this report. However there were many basic and common principles.

Commitment by top management

18.45 Companies with a good safety record are dedicated to the proposition that safety starts with the unfailing commitment of the most senior management, and that of the chief executive officer in particular. They are personally responsible for setting the safety standards for the whole company and for setting the safety philosophy and communicating it to all the workforce. The latter may be expressed in such simple and easily understood concepts as "nothing is so important that it cannot be done safely" or "if we cannot do it safely we won't do it" but underlying those is the belief that safety is a basic element in conducting business and cannot be considered a discrete and separate activity. Safe, prudent working practices and procedures are good business practices.

Creating the safety culture

18.46 It is essential to create a corporate atmosphere or culture in which safety is understood to be, and is accepted as, the number one priority. Management have to communicate the safety philosophy at all times and at all levels within the organisation but most particularly by their everyday decisions and actions in tackling the many issues that arise in operating in the North Sea. Those provide the opportunity for subordinates to see real, practical substance put to the safety philosophy and for exploring the soundness of the safety policy against the realities of operating.

Organising for safety

18.47 To ensure that the safety philosophy becomes a tangible safety programme there must be defined organisational responsibilities for safety; and each part of the organisation has to be set and held accountable for safety objectives. It is essential that from the conceptual design stage of any installation the first objective is to design a safe plant. Thereafter safety has to be a prime objective of on-going operations. Typically the bulk of the responsibility for safety rests with line managers and supervisors, normally backed up by a safety or loss prevention department, which supports and advises line management. Safety objectives have to be built into both short and long term plans, and achievements against those defined objectives have to be part of personnel performance assessment.

Involvement of the workforce

18.48 It is essential that the whole workforce is committed to and involved in safe operations. The first-line supervisors are a key link in achieving that as each is personally responsible for ensuring that all employees, whether the company's own or contractors, are trained to and do work safely and that they not only know how to perform their jobs safely but are convinced that they have a responsibility to do so. Possibly the most visible instrument for the involvement of the workforce in safety is a safety committee system. In Conoco the system involves every member of the platform crew attending a safety committee once per tour of duty, while the Amoco system is based on safety representatives. Both draw no distinction between their own employees and those of contractors. Both companies consider the safety committee system an integral part of managing safety, providing an opportunity for new ideas and new solutions to safety problems to be brought up and a means of passing verbatim and uncensored safety comments up the management line. It also helps reinforce the principle that each employee is responsible for his own safety and that of his fellow workers.

Safety auditing

18.49 Monitoring and auditing the safety process is a critical activity to ensure that any safety programme is being followed. These may be conducted by first line supervisors, managers, safety department staff or personnel from outside the organisation. The requirement for auditing is normally written into company procedures and will encompass the design as well as the operational practices on installations. Audit reports are assessed by management and all recommendations pursued to a conclusion. The chief executive officer will be involved in the processing of the outcome of major audits.

Observations on quality of safety management

18.50 I am convinced from the evidence from both Conoco and Amoco, and indeed from the examination of the background to the Piper disaster, that the quality of safety management by operators is fundamental to offshore safety. No amount of detailed regulations for safety improvements could make up for deficiencies in the way that safety is managed by operators. It therefore is imperative that the quality of safety management should be a component in the regulatory regime. I will return to that issue in Chapter 21, but before doing so I will consider in Chapters 19 and 20 the achievement of the objectives of mitigating the effects of incidents and the securing of safe evacuation or escape and rescue in an emergency situation.

Chapter 19

The Mitigation of Incidents

19.1   The measures which I addressed in the preceding chapter were those required to prevent accidents. In this chapter I turn to measures to mitigate the effects of any accident which may occur. These measures fall under 6 broad headings. Three of these are concerned with minimising the escalation of any leak: (i) the minimisation of hydrocarbon inventory on the platform and in risers and pipelines and isolation of pipelines (paras 19.4-37); (ii) fire and gas detection and emergency shutdown systems (paras 19.38-43); and (iii) fire and explosion protection (paras 19.44-103). The next 2 topics are concerned primarily with the protection of personnel: (iv) temporary safe refuge, or safe haven, escape routes and embarkation points (paras 19.104-175); (v) emergency centres and systems (paras 19.176-194); and the final topic is (vi) pipeline emergency procedures (paras 19.195-197). I conclude with my observations on the mitigation of incidents (paras 19.198-199).

19.2   In the aftermath of Piper there were many calls for there to be requirements for particular hardware solutions and these were echoed in some of the submissions. Examples are subsea isolation valves, blast walls, separate accommodation platforms and enclosed escape routes. I have heard a wide range of evidence on these matters. I will give my views in this chapter on the extent to which it is appropriate for me to make specific recommendations and, where I do not, I will explain how I believe the matter should be handled.

19.3   As I have indicated in Chapter 17, I am in favour of goal-setting regulations and a Safety Case. Some topics, such as fire and explosion protection, may be dealt with both by regulations and by the Safety Case. Broadly speaking, the regulations set goals for the basic design and operation of the system. The Safety Case demonstrates the adequacy of the system in relation to major hazard accidents.

Hydrocarbon inventory, risers and pipelines

19.4   The Piper disaster highlights the importance of the hydrocarbon inventory both on the platform itself and in the pipelines. The scale of the fires was due to the failure first of the Tartan gas riser and later of the other gas risers. In the words of one of the survivors, Mr J M MacDonald: "The Piper did not burn us; it was the other rigs which burnt us." As far as concerns the inventory on the platform, the gas and condensate in the plant appear to have been vented and blown down within a few minutes, but there were significant quantities of oil in B Module, the major sources being the separators and the MOL, which fed the pool fire responsible for the further escalation.

19.5   In this section I will review the evidence on the ways in which hydrocarbon inventory on the platform may be minimised and will consider the prospects for reducing the number of pipelines and risers connected to it and, failing that, for minimising the risk from them, in particular the various types of valve available for shutting off the flow in an emergency.

Minimisation of installation inventory

19.6   Increasingly minimisation of hydrocarbon inventory is being made a design objective for onshore plants. It was a major theme of the Advisory Committee on Major Hazards (ACMH). Such inventory reduction is a specific example of the more general principle of inherently safer design, that is to say, designing hazards out. It is necessarily more difficult in any plant handling fuels. Nevertheless, the principle is valid offshore also and should be applied.

19.7 A well designed venting and blowdown system is able to dispose of most of the inventory of hydrocarbon gas and condensate on the platform to flare within a few minutes. The guidance notes to the Construction and Survey Regulations deal with blowdown and venting of the hydrocarbon inventory. There appears to be no other explicit requirement for, or guidance on, the minimisation of the hydrocarbon inventory on the platform.

19.8 The other main sources of hydrocarbons are the liquid inventories, particularly the oil in the separators and the diesel fuel. The contribution of the oil in the separators to the fires on Piper has already been described; that of the diesel fuel is not known, but may well have been significant. Mr A B Fleishman, Senior Safety Engineer with BP International, in his risk assessment of the Gyda platform found that the diesel fuel gave rise to an appreciable risk. Measures taken to minimise inventory in the design of the Shell Kittiwake platform were described by Mr P A C Doble, Deputy Project Manager. He devoted a section of his account of the design to this topic and described the efforts made to minimise inventory in the separators.

Minimisation of number of risers

19.9 In accordance with the principle of inherently safer design, the first possibility which should be considered in addressing the hazard from the risers is to remove them altogether.

19.10 Some means of exporting the products, oil and/or gas, is unavoidable on a production platform. Such a platform will normally have an oil and/or gas export riser. In many cases platforms have additional risers. Historically, there are a number of reasons for this, including the ease of tying in a new connection on a platform compared with doing so beneath the sea; and the need of some platforms such as Claymore to import gas. Developments in the UKCS, spoken to by Dr Taylor, to which I referred in Chapter 16, mean that there will be an increasing trend to tie in satellite developments to existing platforms by pipeline so that the number of risers will tend to increase. Many of these pipelines will be flow lines, containing 2-phase gas-oil mixtures.

19.11 Evidence on the potential for reducing the risk from risers by keeping their number to a minimum was given by Mr R Willatt, Senior Pipeline Engineer in the Engineering Pipelines Group of BP. The burden of his evidence was that the scope for reducing the proportion of pipelines which were brought on to platforms was limited. Typically additional pipelines were brought to a platform from satellite developments and from remote platforms. The need to tie in a pipeline to other pipelines was one reason. There were available methods of undersea tie-in such as the use of Y and T pieces, but these might involve problems of line diameter, pressure letdown and pigging. Another reason was that for satellite developments the considerable advances made in subsea separation and instrumentation had not yet obviated the need for fluid processing and metering on a parent platform.

19.12 A separate riser platform allowed pipelines to be brought to a main platform which might have limited space or inadequate strength for additional risers. Normally the riser platform would be bridge-linked to the main platform and the pipelines would pass across a pipe bridge. The main platform would be less vulnerable to riser failure and the risers would be less at risk from process incidents. A separate riser platform was, however, a very costly solution. Mr Willatt was unaware of any platform which has the sole function of supporting risers.

Minimisation of risk from risers

19.13 Given that a pipeline is to come to a platform, measures need to be taken to minimise the risk from the riser. Measures available include the design of the riser, location of the riser, fire protection of the riser, and the fitting of valves which will shut off flow. These may include a topsides emergency shutdown valve (ESV) and subsea valves, whether a non-return valve (NRV) or a subsea isolation valve (SSIV).

19.14   Mr Willatt described some of the features to be considered in location of risers, in order both to reduce the frequency, and to limit the consequences of, failure. The riser might be at risk from fire and explosion from process or wellhead areas or other risers, from dropped objects, or from attendant vessels. It might constitute a hazard to the safe haven and the control room and to the process and drilling areas. He advocated that the length of the riser pipe between sea level and the pig trap be kept to a minimum and long horizontal runs avoided.

Fire protection of risers

19.15   Fire protection of a riser may be active fire protection using a water deluge or passive fire protection using a fire resistant coating. Passive fire protection using a fire resistant coating appears attractive because it is less liable to be disabled by the incident itself. Unfortunately, it has the disadvantage that corrosion of the riser pipe may occur under, and be aggravated by, the coating. Mr Ognedal described one of the first major accidents in the North Sea which occurred on 1 November 1975 when a 10 inch riser burst on Ekofisk A, a severe fire followed and 3 men died due to maloperation of a rescue capsule. The failure was caused by corrosion of a section of riser which had been repaired but had not been properly recoated and which was located more or less at sea level. Passive fire protection of risers was explored with several witnesses. The risk from corrosion will depend partly on the type of corrosion which occurs. Some types reduce the thickness of the metal, others weaken the metal which remains; the latter would be particularly insidious. It was Mr Willatt's expectation that the corrosion would tend to be general pitting corrosion, but that the extent of corrosion which might occur between normal inspection intervals could be significant. On the fireproofing of risers Dr R B Gilbert, Chief Engineer on the Nelson Project Team with Shell, was asked whether the state of knowledge was such that it was not known whether the application of fireproofing might make the situation worse or not, and agreed this was a fair statement of the position. Mr A J Adams, Principal Pipeline Inspector with the Safety Directorate, described a joint programme of research commissioned by UKOOA and the DEn at the British Gas Spadeadam site on the ability of coatings used for fireproofing to withstand the erosive effect of jet flames.

19.16   With regard to practice in respect of the fireproofing of risers, Mr Adams stated that it was not common practice. Mr E F Brandie, Safety and Compliance Manager of Chevron (UK) Ltd, knew of no riser with fireproofing in the UKCS. Mr Ognedal stated that since the Ekofisk A accident it has been normal practice in the NCS to consider fireproofing of risers, but he was unable to give even an approximate figure for the proportion which are fireproofed. Mr T Nordgard of Statoil stated that the risers on Gullfaks A are not fireproofed.

Observations on minimisation of inventory and on risers

19.17   The minimisation of the hydrocarbon inventory both on the installation and in the risers and pipelines connected to it should be a design objective and should be a feature of the Safety Case. As regards the former, the Safety Case should address the minimisation of hydrocarbon inventory not only in the main process plant but also in fuel storages such as those for diesel and aviation fuel.

19.18   Emergency disposal of gas and condensate is effected during ESD through the venting and blowdown system. No evidence was heard of any serious deficiency on Piper in the venting and blowdown system, either during the disaster or otherwise. However, it is right to emphasise that this system is a vital part of the arrangements for preventing the fuel inventory from feeding a fire.

19.19   A major role was played in the Piper disaster by a large pool fire. The risk of such a pool fire would be greatly reduced if some method could be found of disposing of the large oil inventories such as those in the separators. This is doubtless not straightforward, but studies should be undertaken to determine whether a practical method can be found.

19.20   Control of the hazards from hydrocarbon risers should be a feature of the Safety Case, but should also be addressed in those regulations dealing with aspects which bear on the problem, including those dealing with the emergency shutdown system and with fire and explosion protection. Possible risks from later, additional risers should also be considered in the Safety Case.

19.21   Studies should be done in support of the aim of minimising pipeline connections to platforms. The development of subsea technology for fluid treatment and metering should be progressed so that there is less need to bring pipelines to platforms.

19.22   The regulatory body should press hard for the resolution of the question of passive fire protection of risers. Passive fire protection of risers is attractive in that it appears less likely to be disabled by the incident itself. There is the risk, however, that corrosion of the riser may occur beneath the coating and actually cause riser failure. There is also some question whether the coatings available will withstand jet flames. Work needs to be done on both these aspects. The aim should be to bring the technology rapidly to the point where either such protection is a reasonably practicable option in a much larger proportion of cases or it is shown that it does not have a significant contribution.

19.23   Active fire protection of risers should not be neglected, but due allowance should be made in any assessment for the possibility that such protection will be disabled by the incident.

Emergency isolation of risers

19.24   Prior to the Piper disaster the isolation requirements for pipelines were those given in the Submarine Pipelines Safety Regulations. Reg 6 requires the provision of effective means of shutting down a controlled pipeline at each of its initial termination points. The inadequacy of these arrangements was revealed when, following Piper, the DEn wrote to operators requesting them to examine their arrangements for pipeline isolation. Analysis of the responses indicated that there were appreciable differences in respect of such valves and 2 principal defects. Some risers had valves which were not true ESVs and needed changes to their actuators, control logic, etc. Although some valves were located near sea level, others were much higher up. Mr Adams agreed that prior to the disaster neither the industry nor the DEn had appreciated the importance of locating these valves low down on the platform. I note that Piper was provided with ESVs on the 3 gas risers, but that one of these, the Claymore line ESV, had only recently been uprated and that the valves were not near sea level but high up on the platform.

19.25 The regulations now made by the DEn, the Emergency Pipe-line Valves Regulations 1989, require that a full ESV be fitted on a riser and that, in effect, this valve be located as near to sea level as practicable. The regulations apply to some 400 existing pipelines. They have resulted in modifications to some 200 pipelines. Of these modifications some 130 involve relocation of the ESVs, some 40 upgrading of the valves to make them fully functional as ESVs and some 30 installation of new valves where the existing valves were not suitable for such upgrading.

Subsea valves

19.26   Another method of isolating a pipeline is the use of a subsea valve. Such a valve needs to be located some distance from the platform, so that it is less at risk from objects dropped from vessels or dragging anchors and so that it is far enough away to ensure that a gas cloud from a rupture on its far side is not ignited from the platform. There is necessarily therefore an appreciable inventory in the pipeline between the valve and the platform. It was emphasised by Dr Gilbert that if riser rupture occurs, a subsea valve cannot prevent a release; it can only mitigate it.

19.27   There is a distinction to be made in subsea valves between NRVs and SSIVs. An NRV may be installed on an export pipeline, but, by its nature, not on an import line. An NRV has the advantage that it responds very rapidly, but it has a number of disadvantages. It may prevent a flow in an export pipeline being reversed for operational reasons such as routine depressurisation, it makes it more difficult to pig the line, it is liable to be damaged by pigging and it will not prevent a small leak.

Subsea isolation valves

19.28   Two studies conducted to assist in deciding whether to install subsea isolation valves (SSIVs) were presented to the Inquiry, one by Dr Gilbert and one by Mr M P Broadribb, Central Safety Engineering Superintendent with BP Exploration. Although the approaches taken in these 2 studies appeared quite different, the BP work comprising a full QRA and the Shell work concentrating on consequence modelling, Mr Gilbert did not admit any fundamental difference. In his case there were no areas of doubt which might make a full QRA necessary. The company had sufficient information from the consequence modelling to make its decision.

19.29   The Shell study highlighted the importance of the criteria used for the integrity of the accommodation and its supporting structure with a jet flame playing on it. For quarters with an A60 wall it was estimated that the air temperature would reach the breathing tolerance limit in about 25-30 minutes, but that fumes generated from the insulation would render the air unbreathable within about 16-17 minutes. The endurance of the quarters was thus taken as 17 minutes. It was estimated that an unprotected supporting structure would fail to support the quarters after about 8 minutes, and this period was thus taken for the endurance of such a structure. The endurance of a fireproofed structural support was estimated as about 60 minutes.

19.30   Another significant point in the Shell study was the effect of the duration as well as the length of the flame. A full bore rupture was not necessarily the worst case. Partial rupture which resulted in a longer duration flame was in some cases a greater threat. In modelling the flames use was made of research into large natural gas jet fires carried out in 1988 by Shell and British Gas at Shell's Thornton Research Centre.

19.31   The Shell study, which covered some 48 gas risers, resulted in the decision to install SSIVs on the 8 risers rupture of which could cause failure of the quarters; and to take other measures in the case of 11 other risers. In 3 cases these measures were to provide shielding for the quarters and in the other cases to review the fireproofing of the structure. The BP study had led to recommendations of about 8 SSIVs on some 5 separate installations.

Subsea valve reliability

19.32   The reliability of all 3 types of valve - ESVs, NRVs and SSIVs - was explored in some detail. For all 3 types the reliability of prime interest here is the probability of giving tight shutoff on demand. The other aspect of reliability is the probability of avoiding spurious action. For ESVs Mr Broadribb quoted a 0.97 reliability for tight shutoff based on a published collection of offshore reliability data. For subsea valves Dr Gilbert stated that the reliability has in the past proved less than satisfactory. He doubted if data on the probability of successful operation on demand of any large population of such valves were available. From the cases he quoted I understood that he was referring mainly to NRVs. As for SSIVs, Mr Broadribb described them as not yet a mature technology. The valves are not commercially available in the full range of sizes and classes. The number of SSIVs installed is not large and the database from which to determine their reliability is therefore small. He considered SSIVs as nowhere near as reliable as the equivalent topsides valves. The value which he used for the reliability of SSIVs was 0.95, which he said was a figure which the industry expects to be able to achieve in the relatively near future.

Subsea isolation valve practicalities

19.33 Dr Gilbert was questioned on the details of the SSIVs which Shell are installing. The installations will consist of a pair of valves at each location. The larger 30 inch valves are among the largest produced; the valves are ''specials''; and the number of suppliers is limited. The time for delivery was estimated as perhaps 15 months and that for completion some 6 months thereafter, but installation would be possible only during the summer. He was not able to give the cost of an SSIV or its installation, but agreed that putting in the SSIVs described was one of the largest single safety bills currently faced by his company and that an estimate for the cost of the best part of £50m seemed reasonable. It was also indicated in the submission by Occidental that the company believes that the technology of SSIVs has progressed to the stage where these valves have a significant contribution to make to safety, though they are not necessary on every pipeline. Such valves have been installed on the hydrocarbon pipelines on the Claymore platform and it is the intention to install them on new platforms. As far as concerns Norwegian practice, it was Mr Ognedal's evidence that there are NRVs installed in, for example, the Statfjord field, but that to date there were no SSIVs. Research by Shell on the reliability of subsea valves was described by Dr Gilbert. A programme to improve valve reliability was started in the early 1980s and is still continuing. The work is concerned particularly with problems of corrosion and maintainability on the seabed.

Observations on emergency isolation and subsea valves

19.34 Hydrocarbon risers, except those which present no threat, should be fitted with a full ESV located as near to the sea level as practicable, taking into account the need to avoid corrosion and to maintain the valve. The DEn has already acted to ensure this through the Emergency Pipe-line Valves Regulations. This action is endorsed.

19.35 The requirement to have an ESV and to locate it near sea level is a specific prescriptive requirement relating to hardware. In general I take the view that prescriptive requirements on hardware are undesirable, but there are exceptions. There are cases where a measure addresses a significant hazard where there is an overwhelmingly clear balance of advantage in its favour; where it is clearly reasonably practicable; and where this situation is likely to pertain for an appreciable period, so that it is inconceivable that it should not be taken. I regard this as such a case. My support for the DEn regulation in this case, therefore, is based on the view that it is appropriate, and not simply on an unwillingness to disturb a set of regulations before they have had a fair trial.

19.36 It was submitted that the Inquiry should also recommend that there should be a requirement for SSIVs. The Trade Union Group and the Piper Disaster Group wished to see essentially default requirements for SSIVs. However, I accept the evidence that there is a wide variety of situations involving risers and that the variation of risk is correspondingly great. This being the case, SSIVs can make a major contribution to safety by reducing the risk from risers, but they are a reasonably practicable solution only in a proportion of cases. The proper approach, therefore, is to determine the need for SSIVs on the basis of the Safety Case.

19.37 Nevertheless, it is my view that the evidence also shows that if progress were made in the technology of SSIVs, there would be a larger proportion of cases where it would be reasonably practicable to use them. It is praiseworthy that some companies have gone ahead and installed SSIVs despite the undoubted difficulties. There remains, however, a chicken and egg situation: installation of SSIVs is held up by lack of information on performance but the latter will be slow in coming unless more valves are installed. There needs, therefore, to be pressure towards the development of more reliable and less costly subsea isolation valves. The aim should be to bring the technology rapidly to a point where such a valve is a reasonably practicable option in a much larger proportion of cases. Since it is the field performance of these devices which matters and since a reasonably large sample is likely to be required, the regulatory body may need to develop unconventional methods of progressing this work. Configurations where there are 2 SSIVs in parallel will permit more frequent test closures. Work should also be done to advance the technology of NRVs.

Fire and gas detection and emergency shutdown systems

19.38 Turning to fire and gas detection and emergency shutdown systems, there are 3 main points which I take from the evidence on these systems on Piper. Firstly, the explosion occurred before signals from the gas detection system had led to either a manual or automatic ESD. Secondly, the ESD of the gas pipelines was not part of the platform ESD, and ESD had to be effected manually for each pipeline separately from the Control Room. Thirdly, some of the ESVs appear not to have closed fully.

19.39 I did not seek evidence in Part 2 on fire and gas (F&G) detection systems, but it is convenient to mention here a particular point made by Mr E F Brandie, Safety and Compliance Manager of Chevron (UK) Ltd and Chairman of the UKOOA Fire Protection Working Group. It concerns infra-red (IR) fire detectors. In the guidance notes to the Fire Fighting Equipment Regulations the fire detection devices referred to are limited to ultra-violet (UV) detectors, but these have proved historically to be prone to spurious trips from welding or from flare radiation with the result that systems are sometimes keyed out. He went so far as to agree that for some applications such systems were not fit for purpose. IR detectors which at one time were prone to spurious trips are now much more reliable in this regard and should be allowed as an alternative to UV devices. He did not, however, recommend a blanket change.

19.40 I have already referred in Chapter 18 to the account given by Mr Ashworth of control and ESD systems.

Observations on fire and gas detection systems

19.41 In the light of the evidence in Part 1 I am not convinced that gas detection systems are making their full contribution to protecting against leaks which may cause serious explosions. In particular, if a leak occurs which warrants an ESD, it is very desirable that this ESD be effected before ignition of the leak occurs, since there is a risk that an explosion will interfere with the smooth execution of the ESD. In the case of Piper the leak ignited quite quickly and it is perhaps debatable whether a gas detection system which gave higher quality information would have made much difference. In other cases it might.

Observations on emergency shutdown systems

19.42 In general, ESD is well covered in the guidance notes to the Construction and Survey Regulations, but I am concerned by the 2 points which I mentioned above. One is the activation of the ESD for the pipelines. There were reasons for the system on Piper in which ESD had to be effected separately for each gas pipeline, since ESD of a pipeline would force an ESD on the connected platform and such forced ESD is generally undesirable. However, the arrangements for the ESD of pipelines are a matter of some importance if the full value of ESVs and SSIVs is to be realised. They should be one of the features considered in the Safety Case.

19.43 The second point concerns the failure of ESVs to close under severe accident conditions, which include fire, explosion and strong vibration. Platform vibration, or shock, caused by the explosion was discussed by Dr Cubbage and was one of the few explanations advanced for the apparent incomplete closure of ESVs on Piper. Work needs to be done to determine the vulnerability of ESVs to severe accident conditions and to enhance their ability to survive such conditions.

Fire and explosion protection

19.44 I turn now to protection against fire and explosion. The initial explosion on Piper knocked out the Control Room and disabled power supplies, communications and the fire-water deluge system, and caused severe vibration which may have affected the ESD system. It generated missiles and led within seconds to further releases and fires, in particular a major fire which gave a massive smoke plume enveloping large parts of the installation and leading within about 20 minutes to the rupture of the Tartan riser. From the start the fire and smoke threatened the accommodation and hindered escape, both from the accommodation and outside, and access to and use of the lifeboats.

19.45 I will give in this section my review of the evidence on protection against both fire and explosion, starting with the latter, but will defer my observations till the end, since I have come to the conclusion that what is needed is an integrated approach.

Explosions in partially confined modules

19.46 The explosion on Piper occurred in a partially confined module and it was on this type of explosion that attention was concentrated. An account of partially confined explosions, or vented explosions, was given by Dr G A Chamberlain, Technical Leader of the Explosion Protection Review Task Force of Shell Expro. His account made clear that such an explosion is a complex process. The over-pressure developed in the explosion derives in the first instance from the volume production of hot gas but there is also a contribution from the effect of flame speed. The pressure is reduced by release of gas through vent apertures and increased by obstructions which cause the flame to accelerate. There is also the possibility of an external explosion, unambiguously confirmed only in 1987, which in turn reacts back on the pressure in the module. The severity of a vented explosion depends on a large number of factors, including the fuel, fuel-air ratio, initial temperature, initial turbulence, location and strength of the ignition source, enclosure size, vent area and obstructed regions.

19.47 There is no fully satisfactory fundamental method of predicting the over-pressure of an explosion within a vented enclosure. There exist empirical equations, but they are generally of limited use; they have usually been derived for empty vessels and tend not to take into account complicating factors such as internal obstacles or external explosions. More useful are computer models and scale model experiments. Computer models include the FLACS code of CMI and the CLICHE code of British Gas, to which I have already referred in Chapter 5. Dr Chamberlain also described another kind of computer model exemplified by the Shell VENTEX code. The model is semi-empirical and is based on extensive experimental work at Thornton Research Centre. The purpose of the code is to provide the engineer with a knowledge of the principal features of a vented explosion and an indication of the extent of the hazard, prior to the use of more fundamental models. These theoretical models may be complemented by scale model experiments.

19.48 Turning to practical applications, Dr Chamberlain described the features which affect the severity of a vented explosion and which the designer should take into account. The explosion severity is minimised if the volume of the enclosure and the extent of the obstacles in it are minimised and the vent area is maximised. Long narrow modules should be avoided. The distance between obstacles should be increased and the blockage ratio decreased. He presented a number of examples illustrating layout features which enhance or reduce the severity of a vented explosion. The principal measures favoured by Dr Chamberlain to mitigate an explosion in a partially vented module were good ventilation and good venting. Essentially the approach advocated was to keep down the over-pressure of the explosion.

19.49 Dr M W Vasey, Manager of Safety Modelling and Offshore Safety at British Gas Midlands Research Station, described work being done by British Gas to investigate the efficacy of water deluge for the suppression of explosions in modules. The concept behind the work was that if water deluge could be activated to drench a flammable gas cloud before it ignited, the strength of the explosion might be greatly diminished. This concept is being tested in experimental work at Spadeadam by British Gas. The work has shown that water deluge can effect an appreciable reduction in explosion over-pressure. There is, however, a significant problem in activation of the deluge, which, to be effective, needs to be operating before ignition occurs. Dr Vasey envisaged that it might be activated by the gas detection system, but acknowledged a difficulty in effecting such activation and suggested as an alternative the possibility of water curtains spaced at about 5m intervals and running continuously. He agreed this would affect the natural ventilation.

Explosion mitigation by venting and other methods

19.50 Dr Chamberlain went on to describe the design options for mitigation of explosions by way of venting and other methods. He prefaced his remarks by emphasising the importance of preventing explosions by eliminating leaks and ignition sources and by dispersing leaks by ventilation. The measures which have been used to mitigate any explosion which does occur are to provide vents to reduce the over-pressure and blast walls to contain it.

19.51 Vent area may be provided in a module by leaving open the ends, by the use of open grating for floors and ceilings, by putting hatches in ceilings and by removing walls or weakening them so they fail at low pressures. There has been some move towards the use at open ends of lightweight weather barriers which both promote natural ventilation and provide a vent area. Where an open grating floor might create problems with spillages, use has been made of lightweight blowout panels sufficient to channel away any spillage but weak enough to come off in the event of an explosion.

19.52 For a new platform, measures which can be taken to mitigate an explosion include the layout of the modules and of the equipment within them. Venting is less effective in long, narrow modules, particularly if the ignition source is far from the open end; short, wide modules are preferable in this regard. The vessels, equipment and pipework may be arranged so as to reduce their effect as obstructions in enhancing an explosion. For an existing platform, where the layout is fixed and the equipment not readily rearranged, venting may be improved by modification of walls, floors and ceilings, as described above. Dr Chamberlain gave examples of retrofitting involving the installation of grated floors and ceilings and of removal or weakening of walls. This general approach was supported by Dr Vasey, who stated that it was his belief that if walls between modules were removed and solid floors replaced by grating, this would greatly improve the effectiveness not only of venting of explosions but of dispersion of leaks. However, he drew attention to the fact that in some cases this may be contrary to regulatory requirements. The allusion was evidently to Reg 11 of the Fire Fighting Equipment Regulations and the associated guidance on reference areas.

19.53 The practical application of methods of mitigation of explosions, and particularly venting, in the design of the Kittiwake platform was described by Mr Doble. Studies were done involving extensive wind tunnel tests to ensure good ventilation to disperse leaks. The segregation of vulnerable features from high risk areas by platform layout provided the first defence against explosion. Explosion modelling was used to assess the risk from these areas. Venting was utilised extensively to minimise explosion over-pressures. The outer sides of the process and wellhead modules were provided with walls of just sufficient area to provide protection from the weather and with gaps top and bottom and weak enough to act as vent panels in the event of a strong explosion. In some areas, particularly the wellheads, use was made of grated decks. Walls in the process area at one end of the platform constituted vent panels. Mr Doble was examined on possible difficulties due to spillages in using grated floors. He agreed there was some problem, particularly with mud spillages in the wellhead area, but did not regard it as a major one, in either the wellhead or the process areas.

Explosion containment by blast walls

19.54   The design of blast walls to contain an explosion was described by Mr A W Van Beek, Head of Offshore Structures Engineering of Shell Expro. He stressed that the designer should seek first to reduce the explosion over-pressure by venting and other measures. For a new platform there was a range of layout and venting options, while for an existing one the venting method which was often practical was the use of open grating.

19.55   As far as concerns existing platforms, Mr Van Beek confirmed that some platforms had only firewalls and not blast walls. He was asked whether such firewalls could be strengthened to protect against blast, but was reluctant to commit himself. He agreed that the measures to strengthen a wall which he described might well mean tearing the whole wall apart. He was asked whether he was aware of any blast wall fitted since the Piper disaster, but did not know of any. Installation of blast walls presented problems of space and weight, which were especially severe for retrofitting of existing platforms. Access might be a further difficulty. With regard to the strength of blast walls on existing platforms he was reluctant to generalise, but stated that from analyses done the strength of a typical blast wall (not a firewall) on an existing platform was of the order of 0.4-0.5 bar. In new designs a typical range of strengths was 0.2-1 bar.

19.56   Comments on blast walls were also made by Mr Brandie. He pointed out that the fire resistance of a wall bore no relation to its blast resistance; a wall with a 1-hour rating might withstand explosion over-pressures better than one with 4 hours resistance. He was asked about combined fire and blast walls; he believed walls approved against both fire and blast existed. He did not know of any installed. He regarded firewalls and blast walls as "different animals".

19.57   Although in the Kittiwake design the prime emphasis is on venting, use is also made of blast walls. Two main blast walls are installed to contain the effects of an explosion, one between the process area and the wellheads; and one between the wellheads and the utilities. Earlier blast walls had been designed to withstand by elastic deformation an over-pressure of 0.3 bar. The blast walls on Kittiwake have been designed using explosion modelling and using an alternative failure criterion based on plastic deformation. Mr Doble stated that it was practice in such modelling to use a worst case ignition source location and he believed the scenario considered had been a module filled with a stoichiometric mixture. For the blowout preventer (BOP) area the original predicted over-pressure was 0.8 bar. In this case there was scope to provide 25% additional vent area by utilising the area under the drilling derrick; and the predicted over-pressure was reduced to 0.6 bar. The blast wall was designed to withstand 0.6 bar. For the separator area the predicted over-pressure was 0.9 bar and no method of improving the venting had been found. It was necessary to design a correspondingly stronger blast wall.

19.58   The use of a blast wall involves the danger that if an over-pressure occurs which the wall is not strong enough to withstand, the wall will disintegrate and give rise to missiles. The higher the pressure at which the disintegration occurs, the greater the energy imparted to the missiles.

Fire prevention and protection

19.59   Fire prevention and protection was spoken to by Mr Brandie in a paper devoted to this and also in his paper on safe haven. He began by listing some of the basic concepts of prevention, mitigation and protection, described active and passive fire protection and went on to develop the argument that currently design against fire was hampered by the regulatory requirements and that the regime should move to one based on fire risk analysis.

Fire risk analysis

19.60   In Mr Brandie's view the principal threat to the platform was major fires. Essentially means existed to cope with lesser fires, but major fires were becoming of growing concern to fire engineers. Increasingly the fire hazard was the subject of a fire risk analysis. This involved in the first place the identification of fire scenarios. The scenarios might be studied using the methods of risk analysis such as fault trees. The risk might then be eliminated. Alternatively, preventive or protective measures might be taken.

Pool fires, jet fires and smoke

19.61   The 2 principal types of fire against which fire protection is required are pool fires and jet fires. In Mr Brandie's view jet fires constituted the greater threat. He felt fairly confident of being able to control any but the largest pool fire; he was less confident about a jet fire. As videos shown to the Inquiry illustrated, a large jet flame may traverse a module. The heat flux from a jet fire tends to be much greater than that from a pool fire. According to Dr Gilbert, the levels of thermal radiation from a non-impinging flame, from an enveloping pool fire and from a riser jet fire were some 100, 250 and 300 kW/m^2, respectively.

19.62   However, as the disaster showed, a large pool fire is also a significant hazard. One measure which can be taken to minimise this is reduction of inventory. Another is sloping the floor under vessels and pipework containing significant inventory so that any liquid oil spill is drained away, and generally taking steps to minimise the areas of potential pool fires. As far as control of a pool fire is concerned, the method used is to smother it with foam inducted through the regular water deluge system.

19.63   As far as Mr Brandie was aware, the minimisation of smoke from platform fires had not been *per se* the subject of much investigation, but measures which minimise or control a pool fire, particularly foam blanketing, would reduce smoke.

Active and passive fire protection

19.64   Ideally fire protection should be a suitable combination of active and passive measures, but in Mr Brandie's experience the adoption of the best technical solution has been hampered by the split of the regulatory requirements for passive and active fire protection between 2 different sets of regulations administered for the DEn by different authorities. Passive fire protection was dealt with in the Construction and Survey Regulations and active fire protection in the Fire Fighting Equipment Regulations. For these the industry had to deal with the certifying authorities and the DoT, respectively. The existing guidance on passive fire protection did not actually prevent the operator from implementing active fire protection in addition but by failing to allow credit for the active fire protection it tended to frustrate the best technical solution. Similarly, guidance on active fire protection tended not to allow credit for passive measures. Moreover, Mr Brandie believed that the volume of detailed guidance, running to some 80 pages, on active fire protection had led to an over-emphasis on this to the detriment of other measures. In his view there were other aspects of fire protection which in some cases were more important, including layout and passive fire protection, but there was much less guidance on these. He stated that in 1986 the UKOOA Fire Protection Work Group had made a strong plea to the DEn for the relevant sections of the 2 guidance notes to these 2 sets of regulations to be amalgamated. However, he saw the draft fourth edition of the Construction and Survey guidance notes as perpetuating the split.

19.65   Although he described it as a "black art" and a developing one, Mr Brandie considered that passive fire protection had an important role to play. It had the great advantage that it gave immediate protection without the need for specific initiation and was less dependent on systems which might be disabled by an explosion such as power supplies. Moreover, it had relatively low maintenance requirements. As against this it provided protection for a limited duration compared with active systems. It might degrade due to weathering and marine environment effects. Its resistance to impact and explosion and to water jets was uncertain. It might conceal or even aggravate corrosion of the surfaces to which it was applied. Overall, however, his company considered passive fire protection sufficiently valuable to have spent some £2m in the last 2 years on refurbishing and upgrading passive fire protection on 3 of its platforms.

Water deluge systems by reference areas

19.66   The Fire Fighting Equipment Regulations require in Reg 11 that an installation should have a water deluge system or monitors, or both. The guidance notes introduce the concept of a "reference area" for application of the water bounded by vertical class A divisions or the edges of the installation; and give a minimum water rate of 12.2 litres/m^2 min. This reference area concept is unique to the UK offshore oil industry. It is not recognised for onshore plants in the UK nor does it occur in the codes of the National Fire Protection Association. The guidance notes state that "the intention is to assess proposals for water protection on an installation by installation basis"; and again that "A water deluge system designed and installed in accord with a suitable standard or code which meets the specification of the general reference area, might be accepted." Despite this Mr Brandie estimated that more than 90% of deluge systems installed offshore followed the guidelines in the guidance notes. He pointed out that active fire protection was typically designed by contractors who found it easiest to adhere to the guidance. The approach had become stereotyped.

19.67   One adverse effect had been on ventilation. Reference areas were basically defined as the complete floor areas of hydrocarbon processing modules and were enclosed by A or H rated firewalls and the edges of the installation. In order to limit the size of reference areas, and hence the water to be delivered, additional firewalls had often been used, so that modules became compartmentalised. This reduced natural ventilation and led to the need for mechanical ventilation and might increase the risk of explosion.

19.68   Another effect had been the installation of massive water deluge and pump systems. The deluge systems used involved a vast number of individual nozzles and associated small bore pipework. The delivery of a uniform water rate of 12.2 litres/m^2 min over the whole reference area led to very high water requirements; and the need to provide additional water to counter the 'shadow' area underneath equipment could almost double the requirements.

19.69   Even so, the systems might be of limited effectiveness against major fires. For protection of individual items of equipment, the water directed straight at the floor was "wasted", though it was a high proportion of the total. On the other hand only very limited pool fires would be extinguished by an application of 12.2 litres/m^2 min and this would require the use of foam induction.

19.70   These deluge systems with their small bore pipework were prone to uneven distribution of the water. Discharge nozzles close to the deluge control valve might be at a pressure up to 3 bar higher than remote nozzles. The systems tended to suffer from severe blockage problems. Nozzles and small pipework were prone to plug. This had led to the use of wet tests to check the state of the system, which tended to compound the problem. Often the systems were too large to be drained, flushed with fresh water or blown dry. Plugging even occurred in the headers. Older systems using galvanised steel were particularly prone to blockage, but despite the use of corrosion resistant materials the problem persisted.

19.71   As far as concerns explosion, the water deluge pipework added significantly to the "clutter" effect and might enhance the over-pressure from an explosion. The systems themselves with their small bore pipework were vulnerable to explosion.

Fire protection systems by scenario-based design

19.72   The alternative approach advocated by Mr Brandie was the use of fire risk analysis and what he called "scenario-based design". This was already practised by companies to varying degrees, but he sought a regulatory regime which would positively encourage it. The scenario-based design approach involved carrying out a fire risk analysis to identify and assess the fire scenarios. Measures to control the risk derived from these scenarios might or might not involve active fire protection. The latter was only one weapon in the fire engineer's armoury. In a given case some combination of measures might be more appropriate which might involve in addition or instead measures such as layout or blowdown or passive fire protection. As far as concerned active fire protection, the method involved protection of specific items such as vessels and equipment rather than blanket protection of areas. On the basis of the fire risk analysis the objectives of the fire protection were then defined; these might be to control or to extinguish the fire or to provide fire exposure cooling. A deluge system was then designed to fulfil these functions.

19.73 The system of nozzles and pipework suitable to a deluge directed to specific items was quite different from that required for area coverage. The nozzles required were fewer but larger and they could be selected to ensure better penetration of even jet flames by the water droplets. Larger bore pipes could be used, with consequent benefits in facilitating fresh water flushing, reducing blockage and imbalance problems, minimising the "clutter" effect and rendering the system less vulnerable to explosion. Mr Brandie expected that with this method the total water requirement would be somewhat less, although averaged out over the total area it would comfortably exceed the standard 12.2 litres/m^2 min. This appeared to mean that the standard rate would be exceeded with all deluges operating, but that in practice even the worst design scenario would not require this.

19.74   Mr Brandie saw a number of advantages in his proposed approach. Major fire hazards were specifically addressed and protection afforded commensurate with the risk. Vulnerable items might be protected with larger quantities of water. Water was conserved and directed to points where it would be most effective. He was questioned on possible disadvantages. He believed that the principal perceived disadvantage was that it would require greater expertise in both the operator and the regulatory body. The current approach made design of active fire protection systems fairly straightforward. He was also asked about possible disadvantages of foregoing a uniform deluge. He believed that the omission of the odd scenario in the hazard identification would not be too serious; the system should cope and there was manual fire-fighting back-up. He considered that the system would be more effective against major fires, both jet and pool fires. Items at risk would be protected with larger quantities of water. The system should also be more effective against minor fires, which would tend to be on items protected by the deluge. In any case minor fires were usually controllable without the deluge. Asked whether he saw any role for the reference area concept, he replied that some situations might well be adequately protected by that approach, but he was opposed to bringing in the scenario-based approach simply as an addition to the existing reference area system.

Fire pump systems

19.75   Existing offshore fire-water systems based on the reference area concept had a very large capacity and required very large fire pumps. The pump capacity was governed by Regs 9 and 11 of the Fire Fighting Equipment Regulations which were commonly understood to mean that each fire pump unit should have the capacity to supply the water requirement for the largest single reference area, which was invariably the wellhead area. Fire-water requirements could exceed 3000 m^3/h. The pump capacity required could be very large. Four or even 5 pump units might have to be installed.

19.76   The requirement for assured power supply to the fire pumps had led to the installation of diesel-driven pumps. It had been the invariable practice, therefore, to install large diesel fire pumps. Historically such pumps had required a high degree of maintenance and had had a poor availability on demand, which in turn had necessitated additional pumps to meet the regulatory requirement. Many of these pumps were now ageing, which compounded the problem. The enclosures needed to house such diesel fire pumps were also large. Mr Brandie quoted an enclosure size of 10m x 8m. It was therefore no easy task to locate such enclosures to achieve segregation.

19.77   A move away from reference areas would alleviate this situation, by reducing the fire-water supply capacity needed and permitting the use of smaller, electrically driven pumps. This would ease problems of location, segregation and protection. Provided the electrical supply was assured, such pumps were highly reliable. There remained, however, the problem of disablement of the electrical supplies by severe accident conditions.

19.78   Mr Brandie was asked about the requirements for fire pump availability given in a DEn letter of 31 May 1989. He agreed that this created an apparent requirement for 100% pump unit availability and that such a requirement was based on a different philosophy from that usually applied in the design of protective systems, such as instrumented protective, or trip, systems, where some unavailability, albeit usually a very small one, is accepted.

Hydrocarbon fire test

19.79   I have already discussed the role of acceptance standards. One of the principal such standards required in fire protection work is a hydrocarbon fire test. However, Mr Brandie stated that there was still no internationally recognised hydrocarbon fire test standard, that this was a problem and that he believed it had been so for some 15 years. The conventional fire test involves putting the assembly in a furnace, heating it up according to a standard time-temperature curve, and determining the times of failure of the assembly and of any insulation. There are 2 principal types of fire test, those for cellulosic, basically wood, fires and those for hydrocarbon fires. These lead to A and H ratings, respectively. Failure in the test is defined in terms of integrity and load bearing capacity and of insulation performance of the assembly.

19.80   Fig 19.1 is that shown by Mr Brandie to illustrate a number of well known time-temperature curves. The BS 476 curve is for cellulosic fires and leads to an A rating; its use for hydrocarbon fires leads to a design with a lower safety margin unless suitable allowance is made. Another somewhat similar curve is the American Society for Testing and Materials (ASTM) E119 curve. The SOLAS curve, referenced in the MODU code, is equivalent to this ASTM curve. The curve shown as "oil company" is commonly referred to as the Mobil curve; and purports to be more representative of hydrocarbon fires. The NPD curve is the Norwegian hydrocarbon fire test curve. The curve marked in the figure with an asterisk is an interim hydrocarbon fire curve proposed by the DEn. The Mobil, NPD and DEn curves lead to hydrocarbon fire test, or H, ratings. The Mobil and NPD curves are both well recognised.

19.81   According to Mr Brandie, the time-temperature curve method is regarded as far from satisfactory. He considered there was a consensus that ideally the test should be based on heat flux. This was confirmed by Mr A R McIntosh, Principal Inspector, who stated that the DEn had commissioned the Fire Research Station to develop a heat flux test. However, there proved to be severe practical difficulties and the work was aborted in 1983. Work on such a test had come to a stop. UKOOA submitted that the appropriate Government department should be required to assist industry in developing a hydrocarbon fire test.

19.82 Mr Brandie was at pains to emphasise that the standard fire test was no guarantee of the behaviour of a structure in a real fire. In his words: "I think one of the major misconceptions in fire protection is the belief that an assembly rating indicates the time the assembly will survive in an actual fire."

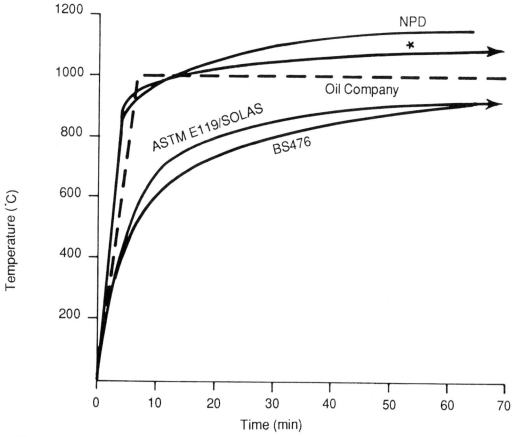

Fig. 19.1 Some principal fire test time-temperature curves. The curve marked with an asterisk is the Department of Energy's interim hydrocarbon fire test curve.

Fire protection in the regulatory regime

19.83 Witnesses led by UKOOA argued that fire protection should become an aspect of the FSA; and the submission of UKOOA was to this effect. UKOOA also submitted that the concept of reference areas and the specification of water rates for the deluge system should be removed from the guidance.

19.84 The evidence of the DEn witnesses was that the Department was to some degree moving in the direction of FSA. The most up-to-date statement of the Department's position was the discussion document on Fire and Explosion Protection, spoken to by Mr McIntosh. The document proposes (Sec 2(a)) that an operator should carry out an FSA of the fire and explosion hazards of the installation and should be able to demonstrate that passive and active fire and explosion protection facilities are sufficient. It also, however, sets out certain specific requirements for all installations (Sec 2(b)). The means of complying with these latter are given in the guidance notes in the form of default criteria, which may need adjustment in the light of the FSA. The document illustrates some of these default criteria. For example, it states that guidance would recommend the use of H120 external boundaries for the accommodation and control stations unless it can be demonstrated that some other standard is sufficient. As another example, it states that guidance would recommend that escape routes between the accommodation and lifeboat embarkation stations close to it should be enclosed and rated to H120 standard unless other standards are demonstrated as satisfactory.

19.85 Questioned on this document, Mr Brandie supported the use of FSA as proposed in Sec 2(a). As far as concerns fire and explosion, he envisaged that there would be a fire risk analysis as part of the FSA. He was opposed, however, to the proposal in Sec 2(b) for specific requirements with default criteria. He interpreted the reference to default criteria as meaning that there would be specific absolute requirements from which the operator would need to apply for an exemption. He thought that the use of FSA and specific requirements was somewhat contradictory. He cited as an illustration of the problem scenario-based design proposals for the Ninian platforms. These proposals were based on fire risk analysis and would supersede the existing system based on reference areas. The company had felt it necessary to put its proposals to representatives of the DEn and the DoT at a single meeting so as to ensure that they were acceptable to both; it still had to "cross the barrier of what the certifying authority might think".

19.86 Mr Brandie considered that the DEn's proposals gave little encouragement to scenario-based design, though it was not actually prohibited. The UKOOA Fire Protection Work Group had held meetings with the DEn over a number of years at which it had pressed for the scenario-based design option, but the DEn had seemed half-hearted. He believed that one reason might be lack of the necessary expertise. Likewise, the perpetuation of the split between active and passive fire protection shown by the draft fourth edition of the Construction and Survey guidance notes suggested that acceptance of a unified approach was still some way off.

Observations on fire and explosion protection

19.87 It is clear to me that prevention of and protection against fire and explosion requires an integrated approach. Design in this area involves balancing a number of factors and making compromises. For example, it is desirable to have good ventilation to disperse any leak which occurs and desirable to prevent fire in one area spreading to another, but the use of additional firewalls for the latter purpose may frustrate the former. There needs, therefore, to be a regulatory framework which facilitates such an integrated approach.

19.88 To this end I have considered 2 options. One is to subsume fire and protection in the Safety Case. The other is to treat it separately by means of its own set of regulations. I have decided to adopt the second option. The fire and explosion hazard will necessarily be a major feature of the Safety Case. However, there are certain features within the Safety Case which have a distinct identity and which will exist on virtually all installations. One of these is fire and explosion protection. It seems sensible to provide for such features by means of goal-setting regulations and thus strengthen the framework of the Safety Case. Furthermore, whereas the emphasis in the Safety Case is on major hazards, these regulations will deal in the usual way with all degrees of hazard.

19.89 The essential requirements for fire and explosion protection should be stated in regulations and should be supported by guidance. Compliance should be demonstrated by reference to a combination of company compliance standards, guidance notes and safety assessment.

19.90 There should be a requirement in the regulations for a fire risk analysis covering both major and lesser hazards. This analysis should involve the identification of the locations where fires may occur; the scenarios of fire and of their escalation; the mitigatory measures available; and the assessment of the hazards and mitigatory measures. The acceptance standards for the design should be developed by the operator.

19.91 The regulations should be framed in such a way as to allow fire protection to be treated as an integrated whole. This means that it should be acceptable for the design options to include the use of active or passive fire protection measures or a combination of the two, and that the design should be assessed on the totality of the measures taken. Here I have in mind the whole range of measures available, including minimisation of hydrocarbon inventory; drainage of spills of flammable liquids; installation layout to segregate vulnerable targets from high risk areas; ventilation to disperse leaks; elimination of ignition sources; systems to give early detection of gas and fire; localisation of fire by fire resistant walls, floors and ceilings; passive measures of fire protection; and active measures to control fire and to cool exposed structures and equipment.

19.92 Likewise, the regulations should be framed in such a way as to allow explosion protection to be treated as an integrated whole, enabling the designer to utilise the whole range of available measures, including installation layout; reduction of over-pressures by equipment layout, venting, and other measures; localisation of explosion effects by blast resistant walls, floors and ceilings; and minimisation of missiles.

19.93 Further, the regulations should be such as to allow protection against explosion to be integrated with protection against fire, so that the designer is free to adopt a design which achieves the best overall compromise between various aspects of fire and explosion protection.

19.94 To be explicit, in order to make best use of advances in knowledge, I believe that the operator should have the freedom to consider designs quite different from those which have pertained historically; specifically, designs involving features such as larger vent areas and more open layouts; more frequent use of combined passive and active fire protection; and water deluge systems which emphasise cooling of equipment. It is not, however, for me to say how this freedom should be used in detail.

19.95 As a general principle, the regulations should be sparing in their use of specific requirements, although, as I have already indicated, I do consider that there are cases where it is inconceivable that there should not be a specific requirement.

19.96 Likewise, as a general principle, the regulations and guidance should be so framed and interpreted as to avoid the creation of default requirements from which variation can be obtained only by means of a lengthy exemption process. For example, requirements for the use of firewalls or blast walls in particular applications and specification of the standards of fire resistance such as A60 or H120 or of blast resistance such as the 0.3 bar criterion, should be used sparingly if at all.

19.97 A fire-water deluge system should be provided to control fires of hydrocarbons which have been released; to cool vessels and equipment containing further fuel which may feed the fire; and to cool fire barriers. The standards for the system should be set by the operator and should cover the function, configuration, capacity and availability of the system and its protection against fire and explosion.

19.98 The regime governing the fire-water deluge system should move towards scenario-based design with no requirement for any particular water deluge rate(s). The scenarios considered in the design of such deluge systems should be comprehensive; they should cover both pool and jet fires; and both small fires which tend to occur more frequently and large fires which occur rarely but which constitute a major threat. I recognise, however, that the current regime and most existing systems are based on the reference area concept and propose that this approach should be retained in guidance as an option, at least in the medium term.

19.99   Similar principles should apply to the fire-water pump system. Current regulations and guidance are onerous, being framed so that they apparently purport to assure zero unavailability. It is indisputable that the availability of these pumps should be very high, but the proper approach is for the operator to set the acceptance standards, in this case the capacity and the availability, and to demonstrate these by FSA.

19.100   The regulations and guidance should promote an approach to the design of fire protection systems which ensures that as far as is reasonably practicable the systems are able to survive severe accident conditions, including fire, explosion and strong vibration. The fire protection systems referred to here include the fire-water deluge system, the fire pump system, and the fire pump startup and changeover controls. The ability of these systems to survive severe accident conditions should also be a feature of the Safety Case.

19.101   The behaviour of a fire barrier under actual accident conditions is inevitably subject to uncertainty, but this is increased by the lack of an internationally recognised hydrocarbon fire test. It is clearly desirable that any test used be realistic. It is equally desirable that problems of devising a test should not prevent or delay the installation of fire barriers which, though perhaps not ideal, nevertheless constitute important safety features. The essential problem is not that of a test *per se* but of the information which the test provides to the designer about the probable behaviour of the fire barrier in real hydrocarbon fire conditions and of the degree of uncertainty the designer can live with, bearing in mind that there is inevitably uncertainty in the fire exposure scenarios. The DEn has already issued an interim test standard. This standard is based on a time-temperature test, the profile of which is broadly intermediate between the widely used Mobil and NPD curves. In the short term the regulatory body should use this standard. It should work with the industry to obtain agreement on how this and other tests should be interpreted for design purposes. If in the view of the regulatory body there exists a need for an improved test, possibly a heat flux test, it should work with the industry to develop one.

19.102   The DEn discussion document on Fire and Explosion Protection is not compatible with the approach just outlined and should be withdrawn.

19.103   The fire risk analysis is one of the measures which the regulatory body should ask operators to undertake forthwith.

Safe haven (or temporary safe refuge), accommodation, escape routes and embarkation points

19.104   I now come to the protection of personnel in the immediate aftermath of a major accident. As the fires on Piper escalated, there was no place which provided protection from the flames and the smoke where they could shelter and try to control the emergency and organise evacuation or escape. Such protection as was provided by the ERQ proved inadequate. Personnel working outside were unable to reach the ERQ or the lifeboats.

19.105   This evidence pointed to the need for there to be on an installation a temporary refuge which provides shelter against fires which may be massive and prolonged and against the associated smoke. There need also to be routes which remain passable long enough for personnel to reach the refuge and to move from the refuge to the embarkation points.

Parties' submissions on safe haven

19.106   The concept of a safe haven was a principal feature in the submissions of the parties. UKOOA proposed that the need for, location and protection of, and facilities in, a safe haven should be a feature of the FSA. A safe haven was part of the submissions of the Trade Union Group, the Piper Disaster Group and the Contractors' Interest, all of whom also made specific proposals for its protection.

Safe haven and temporary safe refuge

19.107   The general assumption was that the safe haven would be the accommodation upgraded as necessary to provide a defined degree of protection. Mr Ognedal stated that the safe haven is part of the total concept of evacuation. It provides a place where persons can remain while either the situation is brought under control or a safe evacuation is organised.

19.108   However, several parties expressed reservations on the use of the term, mainly because it might suggest that there is an area on the installation which can be maintained in a liveable condition for an indefinite period and in all circumstances. UKOOA therefore proposed instead the term "temporary safe refuge" (TSR). I will therefore adopt this latter term. Moreover, for the avoidance of confusion, I will use this term in describing evidence even where the witness referred originally to safe haven.

TSR in the regulatory regime

19.109   There should be a TSR on all manned installations. As I stated in Chapter 17, a central feature of the Safety Case should be a demonstration of the integrity of the TSR in relation to the major hazards of the installation. Thus the TSR imparts structure to the Safety Case. In this section I give further consideration to the TSR and to the associated escape routes, embarkation points and lifeboats, to the construction of the accommodation and to the role of the Safety Case and of regulations in relation to these.

Function of and facilities in TSR

19.110   Mr M J Booth, Head of Operations Safety in the Safety and Environmental Affairs Department at Shell Expro, described the TSR as a place where personnel can muster without being exposed to undue risk. It was a place in which personnel should not simply huddle but should act to assess and control the emergency and prepare evacuation. This is essentially how I see the function of the TSR.

19.111   The concept of a TSR has implications for mustering. The 2 main witnesses who spoke to this topic, Mr Brandie and Mr Booth, said that it was the policy of their companies to muster in the accommodation, but acknowledged that practice differs, another policy being to muster at lifeboat stations. I make no proposals on mustering as such, but clearly any policy on mustering should be compatible with the TSR concept.

19.112   Assessment and control of the emergency from the TSR requires the availability within it of the necessary facilities. I consider this aspect in para 19.176 *et seq.*

Endurance of the TSR

19.113   There was general agreement that the basic approach to the design of the TSR should be to specify an endurance time for occupancy and then to identify the hazards to which it may be exposed; to define agreed scenarios which it is to be designed to withstand; and to perform a risk assessment to confirm the design.

19.114   Two general types of consideration were put forward as governing the choice of endurance time, namely the exhaustion of the threat and the time to arrange evacuation. For example, Mr Brandie referred to the need to allow the platform inventory to exhaust itself; Mr Booth to the need to allow any event on the platform to subside. The risk assessments showed that the safety measures taken obviated the need for a longer period. The riser hazard, for example, was handled by limitation of the risk rather than extension of the endurance time.

19.115   The other consideration determining the endurance is the time to effect evacuation. Here the lead times for helicopter evacuation will tend to give long endurance time requirements. Mr Booth disagreed with the suggestion that the endurance should be such as to allow for evacuation by helicopter.

19.116   Mr Booth stated that the endurance time used by his company was 60 minutes. It was the period necessary rather than the limit of what was technically achievable. Mr Brandie also made reference to an endurance time of 60 minutes, but was reluctant to give a firm figure.

19.117   As far as concerns the practicality of particular endurances, Mr Booth said that risk assessments had been carried out on all his company's platforms. The 60 minutes period was seen as the practical figure to aim for. He supported the use of H120 firewalls where the risk assessment showed that this was necessary, but he considered that even with H120 firewalls it was difficult to assure occupancy for more than 60 minutes and that it was realistic to adopt this figure. He regarded this as consistent with the Norwegian approach, described below.

19.118   It is clear that there are a number of factors which may limit quite severely the period for which the TSR is occupiable. Factors mentioned by Mr Booth included heat, smoke, combustion products, toxic fumes and disintegration. Similar factors underlie the endurance criteria used by Dr Gilbert in his study of SSIVs, which I have already described.

19.119   The endurance set for the TSR in the Norwegian system, which is 2 hours, was spoken to by Mr Ognedal and Mr Nordgard. This comprises one hour for collecting personnel into the TSR and one hour for effecting evacuation. While this time may be to a degree arbitrary, it has some basis in evacuation trials carried out on platforms. The time for blowdown of the hydrocarbon inventory on the platform is also a factor. Mr Nordgard stated that this 2 hour period was given in regulations and in guidance. However, while the Guidelines for Safety Evaluation of Platform Conceptual Design set a time of one hour for the availability of at least one escape route against a design accidental event, there was no corresponding specified period for the shelter area, or TSR. Asked directly where the 2 hour period was actually stated, Mr Ognedal said that it was referred to in terms of the H120 firewall in the Regulation for Production and Auxiliary Systems 1976. Para 6.5.2 of these regulations stated that as a minimum the outer surfaces of the living quarters which might be subjected to a hydrocarbon fire should be protected to H120 standard. He indicated that the NPD had made no decision on whether to make any change to the endurance required. With regard to the endurance achievable, Mr Nordgard was of the view that it was not much more difficult to design for an endurance of 2 hours than for one of one hour and that there was no great benefit in specifying the shorter period. In this context he stated that the design accidental events did not include riser failure; such failures were dealt with by reducing the frequency.

19.120   The DEn discussion document on Fire and Explosion Protection states that the TSR should remain viable for at least 2 hours unless demonstrated otherwise by FSA. Mr McIntosh agreed that this figure was somewhat arbitrary. He said it was related to the proposal that the accommodation should be protected by H120 firewalls unless the FSA showed otherwise.

Protection of accommodation against external fire

19.121   Two principal threats to the TSR are external fire and smoke. The construction of the accommodation is one of the items covered in the Construction and Survey

Regulations. The associated guidance notes give guidance on its protection. The third edition states that the bulkheads separating the accommodation from the wellhead and process areas should be to A60 standard or to a standard providing equivalent protection. The draft fourth edition states that the control stations and the accommodation, as a TSR, should have fire durability commensurate with the possible exposure and reiterates the requirement that boundaries between the accommodation and the wellheads and process areas should be to A60 standard. It adds, however, that where the risk of a hydrocarbon fire exists, it should be assumed in the absence of other information that all external boundaries require H120 protection, unless some other level of protection is shown to be appropriate by reason of the likely extent, duration and severity of the fire exposure.

19.122   It was Mr Brandie's evidence that the general practice in the British sector was to put an A60 division on the side of the accommodation facing the hydrocarbon risk and in a number of cases to continue around the sides and inside, if there was a possibility of exposure. On some platforms this passive protection was complemented by water drench systems. The accommodation protection taken by Dr Gilbert as typical for his company's platforms was A60.

19.123   The Norwegian requirements for fire protection were described by Mr Nordgard and Mr Ognedal. They referred to the Regulation for Production and Auxiliary Systems, which required that outer surfaces of the accommodation which might be exposed to a hydrocarbon fire should be protected to H120 standard and those which might be exposed to fire from other areas should be protected to A60 standard. The fire exposure, and hence the protection required, was obtained from the QRA.

19.124   I have already mentioned, in the context of the hydrocarbon fire test, the point made by Mr Brandie that a particular nominal rating of a firewall was no guarantee that the wall would exhibit that degree of endurance in a real fire. An essentially similar point was made by Mr McIntosh when he said that there was nothing sacrosanct about H120 protection; it was to some extent arbitrary. It was not certain that such protection would necessarily last 2 hours. It represented an improvement on A60 protection rather than an absolute level of protection. Likewise, Mr Booth said he would support the use of H120 protection where FSA showed it to be necessary, but as a means of achieving an endurance time of 60 minutes.

19.125   This evidence indicates to me that the proper approach is to define the endurance required and hence the necessary degree of protection rather than to specify the means in terms of firewalls of a particular rating and that the way to do this is through the Safety Case.

Protection of accommodation against smoke

19.126   The endurance of the TSR will be determined in large part by the breathability of the air within it. This topic was addressed by Mr G A Dalzell, a Fire and Safety Engineer with BP International. The principal factors which might render air in the accommodation unbreathable were heat, smoke and toxic fumes. Heat transfer through the external walls would heat the air and heating of the walls may produce toxic vapours, but he estimated that both effects would be delayed for one to 2 hours, depending on the firewall rating and the fire exposure. Carbon dioxide build-up should not be significant within the 2 hours. Nor should oxygen depletion by the occupants. The main problem addressed by Mr Dalzell was therefore smoke.

19.127   Mr Dalzell reviewed the main potential sources of fuel for smoke-generating fires, outlined some of the scenarios for such fires, and described the use of wind tunnel testing to investigate smoke movement for design purposes. The first lines of defence in reducing the risk of smoke ingress into the accommodation were the prevention of and reduction of scale of fires; orientation of the installation so that the prevailing wind blew the smoke away from the accommodation; layout which segregated the accommodation from areas where smoke generation might occur; and positioning of the accommodation at a low level so that at least its lower part was below the main sources of smoke.

19.128 Weak points on the accommodation through which smoke might enter included penetrations through the external walls, doors, windows, and ventilation air intakes and exhausts. Mr Dalzell distinguished 4 classes of door: main entrance doors; emergency entrance doors; emergency exit doors, or escape doors, otherwise known as crash doors; and evacuation doors. The main and emergency entrance doors needed to be on the front of the module, facing the rest of the platform, and hence were more vulnerable. The evacuation doors should be at the back close to the lifeboats. The escape doors, intended for escape from an internal fire rather than evacuation, were commonly put at the end of corridors; they tend to be fairly numerous. Essentially his proposals for doors were that escape doors should be kept shut and main entrance and evacuation doors provided with air locks. He recognised that escape doors might be used and left open, thus letting smoke in. He believed that mustering at low level and muster discipline should minimise the problem, but agreed that self-closing doors were both desirable and practical and saw some merit in break-glass panic bolts. He considered that escape doors did not need protection by air locks. He did advocate air locks for the main entrance doors, where they had traditionally been fitted, and for evacuation doors. Air locks served to conserve air and maintain positive pressurisation. They could be defeated, however, if a continuous stream of people passed through, so keeping the doors open, and needed therefore to be large enough, say 3-4m, to hold 6-10 people. Few accommodation modules were fitted with windows, though more recent designs might have them on the rear wall. There was a problem in obtaining windows rated for hydrocarbon fires. Mr Dalzell said that windows rated A60 were available and he believed some had been tested to H60 rating. He suggested there might be small strategic observation windows near doors and agreed the problem of fire rating might be overcome by fitting such windows with small covers.

19.129 Smoke could be prevented from entering if there was a positive pressure in the accommodation. This pressure was maintained by the ventilation system, of which there were basically 2 main types. Both consisted of inlet fans, ducting and exhaust fans. A forced ventilation system was the basic type. It might give a high degree of protection against smoke ingress if optimised for this, but it was not designed to maintain positive pressure and some rooms might be at negative pressure. The system was balanced for one set of wind conditions and it could compensate for other conditions. It was also liable to deterioration and needed careful maintenance to achieve optimum performance in excluding smoke. More modern platforms might be fitted with a positive pressure ventilation system, essentially a refinement of forced ventilation, which was more flexible and gave closer control of pressure, typically maintaining 6-12 mm water gauge. Ventilation systems were not classed as emergency systems and generally were not powered from the emergency power supply (EPS). However, on some platforms, for reasons of commissioning or maintenance, the system could take power from the EPS. Some platforms had one inlet and one exhaust fan on the EPS. In general Mr Dalzell was opposed to running the ventilation from the EPS, being reluctant to risk jeopardising the other emergency functions. If the ventilation inlet fans were lost, positive pressure would be maintained only for a few minutes. It would not be maintained if a door was left open, though for a period the air flow would be outwards, reducing smoke ingress. The period during which smoke was excluded could be maximised by the use of air locks on doors and of dampers on the ventilation ducts.

19.130 The ventilation air intakes were a weak point through which noxious gases might be drawn into the accommodation. Prevention involved shutting off the ventilation and closing dampers in the intakes. The extent of provision for automatic shutdown of the ventilation system on detection of fire, gas or smoke was unclear from Mr Dalzell's evidence, but he described in some detail the arrangements for closure of the inlet dampers. For these only remote manual closure and automatic closure on heat detection were universally provided, but it was common in addition to have automatic closure on detection of flammable gas. There might also be automatic closure on detection of a particular toxic gas such as hydrogen sulphide. The arrangements for exclusion of gas and of smoke were similar and in many cases the former already exist, so that it was not difficult to add the latter. Some platforms had smoke detectors on the air inlets prior to Piper; most which did not, had since fitted them. Mr Dalzell considered that reliance on manual closure of the ventilation intake dampers was not appropriate; closure should be automatic. Apart from closure on heat, gas and smoke he favoured closure on loss of power but not on loss of positive pressure, since the latter could be caused by an open door. The vulnerability of the ventilation air intakes to smoke ingress could be reduced by positioning them low down, below the level of most smoke plumes. However, the location must take account of exhaust fumes from platform sources and from vessels. Mr Dalzell conceded that it might be possible to provide emergency intakes, but foresaw a possible problem with changeover.

19.131   Mr Dalzell recommended that all installations should have a smoke ingress assessment of the accommodation module; automatic shutdown of ventilation and closure of intake dampers on smoke detection; and air locks on main entrance and evacuation doors. Relocation of air intakes and doors to minimise their vulnerability to smoke should be considered and also provision of observation windows. Forced ventilation systems of older platforms should be reassessed to improve maintenance of positive pressure and new installations should have positive pressure ventilation systems. For new designs smoke movement should be assessed by wind tunnel testing and smoke ingress should be a factor considered in positioning of the accommodation.

TSR as a citadel within accommodation

19.132   For longer term refuge Mr Brandie envisaged the use within the primary protected areas of a secondary protected area, a "box within a box", and instanced the application of this on the Ninian installations. This is effectively a citadel, although he did not use that word.

Upgrading of accommodation to TSR on existing platforms

19.133   The practicability of upgrading the accommodation to a TSR on existing platforms is clearly of prime importance. Mr Brandie described some of the measures which may be taken on an existing installation. These included uprating A60 firewalls to H rating; use of combined passive and active fire protection; installation of radiation screens; removal or reduction of close proximity hazards; enhancement of structural protection; and major incident prevention. He also drew attention to some of the problems in uprating existing A60 protection to H rating which may make such uprating impractical. He agreed that the problems were common to both A and H class protection.

Additional refuges

19.134   It was recognised that at least on some existing large platforms there might be a need for temporary safe refuges additional to the main TSR for personnel who would need to muster elsewhere. Mr Booth instanced the drill crew, who have to make the wells safe. This was a point which Shell were still considering.

Bridge-linked accommodation platforms and flotels

19.135   One of the principal measures canvassed after Piper has been the provision of accommodation separate from the main production platform, typically in the form of a separate, bridge-linked quarters platform or flotel.

19.136   A generalised study of the comparative risks of different platform configurations, originally commissioned in August 1988 by the DEn from Technica and the

Offshore Certification Bureau, was described by Mr J R Spouge of Technica. The cases considered were: case A: Base case design: a base case platform, representative of design practice in the UK Central and Northern North Sea before systematic consideration was given to the safety implications of topsides layout; case B: Modern design: a modern equivalent of the base case, characteristic of recent design practice in which the topsides layout is heavily influenced by safety considerations; case C: Bridge-linked flotel: the base case with accommodation on an adjacent flotel linked by a bridge; case D: Helicopter-linked flotel: the base case platform with accommodation on a nearby flotel linked by helicopter; case E: Smaller capacity platform: a smaller, 4-legged platform, representative of a modern trend towards lift-installed jackets with cantilevered, integrated decks; case F: Bridge-linked quarters platform: the base case platform with accommodation on a separate quarters jacket platform linked by a bridge. These configurations are illustrated in Fig 19.2. Cases A-E were specified by the DEn and case F for the Inquiry. Initially the aim was to determine the frequency of impairment of structure, accommodation and escape routes by residual accidental events along the lines of a Norwegian CSE, but this was later extended to determining the average annual fatalities from high fatality accidents (10 or more deaths) throughout the drilling and production phases.

19.137   The average annual fatalities in high fatality accidents for the 6 cases are shown in Table 19.1. The study showed that the average annual fatality risk for the modern design, case B, and for the bridge-linked platform, case F, was about a third of that for the older platform, case A, and that for the bridge-linked flotel about a half, while that for the helicopter linked flotel, case D, was about half as much again, due largely to the contribution of helicopter accidents. The smaller capacity platform, case E, had a risk about one ninth, but also a production rate one quarter, of the base case. A sensitivity analysis had been performed but it remained true that the risks of case B were 50-75% lower than those of case A and that the risks of cases B and F were within 20% of each other.

19.138   The reduction in risk as between cases A and B was due mainly to the stronger jacket on the latter; other significant features were topsides layout, firewalls and riser protection. The reduction as between case A and case F, on the other hand, was due largely to reduction of exposure of personnel. In case F the office staff, who numbered 35, were located on the quarters platform so that the proportion of personnel remaining on the main platform was reduced to 25% of the base case. This effect of the location of office staff was a significant result of the study. The benefit was not available in case C, because the flotel had to stand off in rough weather and so office staff had to be accommodated on the main platform.

19.139   Of the measures considered in the study for reducing the risk of an existing installation, Mr Spouge stated that the provision of a bridge-linked quarters platform was the most effective. The study did not, however, address other measures such as relocation of ESVs nearer to sea level or the installation of SSIVs. On the practicality of a bridge-linked quarters platform, he stated that such arrangements existed in the southern sector of the UKCS and in the Norwegian sector of the North Sea. In the northern sector of the UKCS he was unaware of any dedicated quarters platform, though he did know of one 2-platform concept where one platform housed the quarters and some other facilities.

19.140   Mr Spouge was careful to point out the limitations of the study, which was a generalised ranking exercise, and agreed that for a particular site it would be necessary to do a specific study. He stated that the same methodology was in fact being used by operators to assess options, especially in respect of accommodation, ESVs and SSIVs.

19.141   Risk assessments described by Mr Tveit implied a reduction in risk as between an older and a newer platform of about 4, but he was unable to say how typical these figures were.

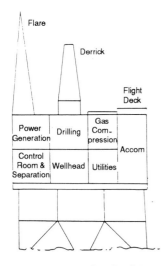

CASE A : BASE CASE

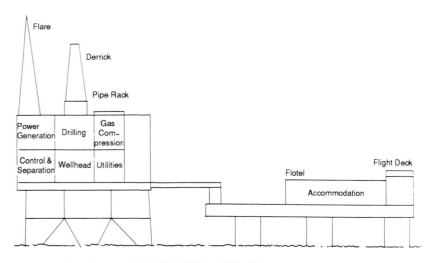

CASE C : PLATFORM AND BRIDGE LINKED FLOTEL

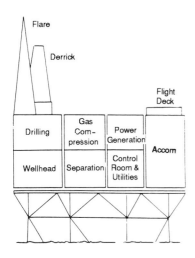

CASE B : MODERN DESIGN

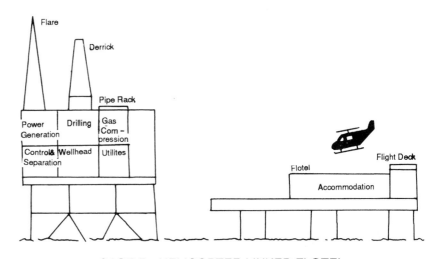

CASE D : HELICOPTER LINKED FLOTEL

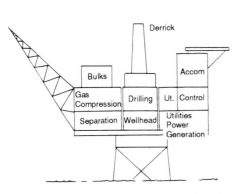

CASE E : SMALLER CAPACITY
PLATFORM

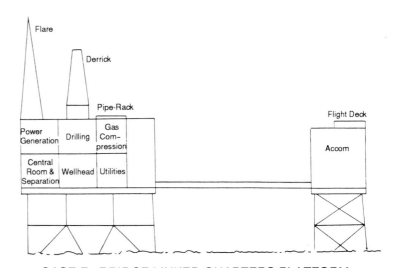

CASE F : BRIDGE LINKED QUARTERS PLATFORM

Fig. 19.2 Installation configurations studied in the generalised quantitative risk assessment conducted by Technica Ltd: (a) base case; (b) modern design; (c) bridge-linked flotel; (d) helicopter-linked flotel; (e) smaller capacity platform design; and (f) bridge-linked quarters platform.

327

19.142   The only example of a decision to install a bridge-linked quarters platform about which I heard evidence was Mr Ferrow's evidence that Conoco had decided to locate the accommodation for their Southern Basin Gas Development, or V Field, on a separate platform. The decision, in early 1984, was made at a very early conceptual stage, and before the major hazard review.

19.143   The Trade Union Group and the Piper Disaster Group submitted in effect that there should be a default requirement for quarters separated from the main production platform, the former advocating a separate bridge-linked accommodation platform and the latter separate accommodation, either fixed or floating. UKOOA disagreed and submitted instead that the location of the accommodation should be a matter to be considered in the FSA.

Escape routes to TSR

19.144   A necessary complement to a TSR is escape routes to and from it. On Piper within minutes of the initial explosion virtually all the escape routes provided to the accommodation became impassable and the area on the north face of the ERQ was engulfed in fire and smoke. Evidence on escape routes was given by Mr Booth, who said that since Piper perceptions had changed, the magnitude of the hazard was better appreciated and companies had been conducting reviews of escape routes on their platforms.

19.145   The thrust of Mr Booth's evidence was that escape routes should generally be the normal routes which were used in moving around the platform and which thus had the advantage of familiarity. Passability should be ensured by layout rather than by protection. On existing platforms careful consideration had gone into the design of routes, but this had related primarily to location, path and number of routes and detailed design features rather than to protection against fire, smoke and falling debris. He listed certain principles of escape route design. Escape routes should be as direct as possible. They should lead from internal areas to an external escape route. Primary escape routes should be located wherever practicable external to the modules. He was in favour of permanent walkways around the perimeter of the platform at all levels. Escape routes should not be routed through areas of increased hazard nor past explosion vents or relief walls. Given that the escape routes were afforded protection by their position, there was generally only limited need for other types of protection.

19.146   Conditions for the passability of escape routes are given in Annex 9 of the Petrie Final Report, a report by Technica, quoted by Mr Booth. The report also refers to the criterion that escape routes should remain passable for 30 minutes. He considered this figure reasonable.

Escape routes from TSR to embarkation points

19.147   As far as concerns the escape routes to the embarkation points Mr Booth envisaged that the embarkation points might be in a Protected Area, a term used in Shell to denote an area in the lee of the accommodation and sheltered from the fire hazard. Thus if the wellhead and process areas were on the south end and the accommodation on the north end of the platform, the Protected Area would be at the north face of the accommodation, on the outside. He envisaged that there would be sufficient lifeboats at this point to take everyone in the accommodation. In answer to the point that the area on the north face of Piper did not in fact provide protection for evacuation by lifeboats but was engulfed by smoke within minutes, he replied that a great deal could be done to reduce the risk and that too much should not be taken from this one case.

19.148   He advocated that the escape routes from the TSR to the embarkation points should be short so that movement through them was swift. He envisaged that such escape routes would often run through the Protected Area. They could if necessary be provided with protection such as radiation screens, structures to prevent falling debris and even firewalls, but he stressed that it was important that the OIM should not lose contact with people going to the lifeboats. This would be the probable result of making the escape route an H120 rated tunnel. He agreed that the escape routes from the TSR should survive so that they were usable by personnel at the end of the endurance period of the TSR.

Protection of escape routes and embarkation points

19.149   The essential requirement for an escape route was that it should be passable. It was preferable to achieve this by layout. Otherwise, it might be necessary to protect either the escape route itself or the people passing along it, or both. The protection required was principally protection against heat and smoke. Given the choice, Mr Booth expressed a strong preference for protection of the escape routes rather than reliance on personal protective equipment.

19.150   The main protection which he described was the use of anti-radiation mesh, usually known as radiation screens. Standard double-sided heat shields would provide protection against fire; they might be complemented by a water spray. Water sprays might also be provided on the escape routes themselves; he referred to the relief provided to people on Piper by the water spray from the *Tharos*. He was strongly opposed to the proposal in the DEn discussion document on Fire and Explosion Protection that escape routes might be protected by an enclosure, or tunnel. He said it "confirmed his worst fears". He considered in effect that such tunnels would actually cause safety problems. There would be problems of access to the tunnel, of its ventilation and lighting, of possible loss of integrity, of ingress of smoke, and of disorientation and possibly panic among people using it. He referred to the 1987 fire at Kings Cross Underground Station as an illustration of some of the problems.

19.151   Escape routes needed to be provided with lighting which would function in an emergency. Mr Booth suggested the use of high intensity emergency lighting with battery back-up. He also referred to the use of photoluminescent signs as suggested by survivors of Piper. He agreed that lighting along the floor of the walkways, as in aircraft, would be helpful and that this might best be provided by photoluminescent strips.

19.152   Mr Booth made no proposals for the protection of escape routes from smoke and, while in general preferring to protect the escape route, he agreed that consideration should be given to smoke-hoods and also to lightweight BA sets to allow people to use the escape routes in the face of smoke. He was less disposed to rely for protection against heat on the use of fire-suits, which were cumbersome.

19.153   The protection which he described for the embarkation points was broadly similar to that which he proposed for the escape routes. He was opposed to enclosed embarkation points and concerned that protection should be provided only where it was shown to be necessary. He envisaged that embarkation points would be open to the sea but with radiation screens inboard where appropriate.

Protection of lifeboats

19.154   The lifeboats needed to survive until it was safe for personnel to leave the TSR and go to them to evacuate the platform. The endurance required of the lifeboats was therefore related to that specified for the TSR. Mr Booth thought that it might be necessary to relocate or protect the lifeboats but believed this would not be necessary in most cases. He saw some merit in the suggestion that the lifeboats might be located within the TSR, though there were potential problems such as doors jamming. He did not think such an arrangement favoured free-fall lifeboats particularly.

Escape routes and embarkation points on existing platforms

19.155 Mr Booth outlined a number of measures which could be taken to uprate the escape routes on existing platforms, prefacing his proposals by stressing that account should be taken of the original design philosophy. Steps which he mentioned included measures to reduce the risk from risers, relocation of lifeboats, provision of additional escape doors in the TSR, and protection of routes. He also suggested the provision of smoke-hoods and BA sets.

Observations on TSR, accommodation, escape routes and embarkation points

19.156 The Piper disaster demonstrates that there is a clear need on a platform for a TSR. The industry has recognised this. Companies are already acting to ensure that the main accommodation is protected to a standard such that it constitutes a TSR. There are a number of reasons why the accommodation is the logical choice. Usually it is located furthest from the more hazardous activities, is often the only suitable space in which to assemble large numbers of personnel and at any given time contains a considerable proportion of those personnel. The TSR is therefore taken here to mean the main accommodation.

19.157 I have already proposed in Chapter 17 that the TSR should be a central feature of the Safety Case. The operator should specify the function, the endurance and other acceptance standards for the TSR and should demonstrate by QRA that it has provided one which meets those standards.

19.158 The acceptance standards for the TSR will be of 3 types. The first is the risk criteria, including one for the frequency of loss of integrity of the TSR. The risk criteria should follow the ALARP principle. The second type is the endurance times. The third type is the standards defining loss of integrity. Formulation of these will involve defining the function which the TSR is to perform, the conditions which constitute integrity, the endurance time for which these conditions are to be maintained and the events which may cause these conditions to be violated. The endurance specified for the TSR will determine which hazardous events are residual accidental events which the TSR is not designed to survive. All types of acceptance standard should be specified by the operator.

19.159 However, initially at least, the regulatory body should set minimum standards for the main risk criterion, the frequency of loss of integrity of the TSR. Further, initially at least, the regulatory body should set a minimum endurance time for the TSR. I have weighed carefully the arguments for and against this. The argument against is basically the general one that this is a matter which is best handled as part of the operator's own FSA. The argument for is that the choice of the risk criterion and the endurance necessarily involves a degree of judgement and is to some extent arbitrary; that any gain in flexibility is outweighed by the introduction of a point of probably rather sterile contention; that there is no detriment to safety; and that it puts the Safety Case on a firmer basis.

19.160 The hazards to which the TSR may be subjected should be identified. The hazardous events should be classified as design accidental events and residual accidental events. The TSR should be designed to survive design accidental events. It need not be designed to withstand residual accidental events, though if put to the test it may well survive some of them. The relation between the risk criterion and the endurance time should be seen in this light. For many scenarios the operator will find it more effective to meet the standard by reducing the risks, especially those of fire and explosion, rather than by providing the TSR with extreme levels of protection. This is so particularly in relation to risks from risers.

19.161 It may not be necessary that the whole accommodation module be nominated as the TSR. It should be an option that the TSR should be a limited, protected area,

330

or citadel, within the module. However, since this is uncharted territory, the approach should be a cautious one. In any event, a design based on such a TSR should meet the full requirements for the TSR and escape routes.

19.162   It may be that in some cases the Safety Case will show that the requirements for the TSR can be met only by the use of accommodation separate from the main production facilities such as on a bridge-linked platform. As I have already indicated, it is my view that decisions of this sort be made in the light of the Safety Case.

19.163   The conditions for the integrity of the accommodation are crucial to the risk assessment for the TSR but it is clear that the criteria currently available are rather crude. There is a need for models of the development of the air conditions in an accommodation module and for criteria against which the results from such models can be assessed. The models should address high temperature due to heat transfer through the walls, smoke due to smoke ingress, and toxic fumes from heated fire insulation and any other likely sources. The endurance time used by Dr Gilbert for living quarters exposed to fire, based on build-up of toxic fumes, was relatively short, only 17 minutes. On the other hand Mr Dalzell did not envisage toxic fumes being a problem for 1-2 hours. Unless these matters are sorted out, they are likely to create difficulties in assessing the integrity of the TSR and hence in the development of the Safety Case.

19.164   Staying with the Safety Case, the TSR should be complemented by escape routes to and from it and by embarkation points and lifeboats. These should be treated along with the TSR as central features of the Safety Case. For each location on the platform at least one escape route to the TSR should be passable for a defined endurance time against design accidental events. Likewise, the escape routes from the TSR to the embarkation points, the embarkation points themselves and the lifeboats should each have a defined endurance against design accident events. These endurances should be defined by the operator.

19.165   For existing installations any requirement for upgrading of the accommodation, or TSR, escape routes, embarkation points and lifeboat protection should be decided on the basis of the Safety Case.

19.166   As the TSR, the accommodation will be dealt with in the Safety Case, but it is also a proper subject for regulations on construction. These regulations should set goals for the design of the accommodation and may also contain specific requirements.

19.167   Before considering the protection of the accommodation, I wish first to comment on the loss of integrity of the main accommodation module on Piper, the ERQ, which I described in Chapter 8, and in particular on the special features of this case. The ERQ was protected on the face nearest the fires, the southern face, by other structures. This must have reduced the extent of generation of toxic gases from the module walls which in other circumstances might be much more serious. On the other hand the ERQ was actually breached by the fire. Flames entered in at least 3 places; at the doorway from the LQW, the southern external door to the helideck and in the north-west corner cabins, notably cabin C1.

19.168   As far as concerns the entry of smoke, the evidence is that the ventilation air intakes were not a major route for smoke, but this appears to have been due largely to the fact that loss of power stopped the ventilation fans. There were no smoke detectors to shut the dampers on the air intakes. But for the fortunate chance of this stoppage large volumes of smoke would probably have been sucked in until the dampers closed on high temperature. There was, however, gross ingress of smoke and hot gases through the doorway from the LQW. The southern external door was a second major route for smoke; it was not possible to say which was the more significant. Smoke also entered through broken windows in the cabins on the north-west corner. It may also have entered through other doors and windows.

19.169   Open fire doors allowed smoke to spread within the ERQ. The door between the reception area on D Deck and the stairwell, which was hooked open, attracted most attention, but smoke also spread through open doors along the north corridor of C Deck and into cabins C1 and C11. On the other hand where fire doors were closed, they were effective. This was the case with the door between the passage from reception and the dining area, the other doors off the stairwell and the doors to other cabins on C Deck north corridor. The reception area on D Deck was both a general thoroughfare and an emergency control centre. It was no doubt for this reason that the door to the stairwell was hooked open. Another major route for spread of smoke within the ERQ was through the ceiling voids. This spread was prevented, however, where walls extended through the ceiling or where there were cavity barriers.

19.170   Regulations on construction should include among the goals for the design of the accommodation protection against external fire, exclusion of smoke, prevention of smoke movement and maintenance of breathable air. They should allow an integrated approach to the achievement of these goals which covers the external firewalls, the internal construction, the doors and the ventilation system.

19.171   The need for an integrated approach is illustrated by the ventilation system. It is clearly essential that smoke should not be sucked into the accommodation through the ventilation intakes. On the other hand, positive pressure maintained by the ventilation system allows the use of air locks to prevent smoke entering through main entrance and evacuation doors. The power supply for the ventilation system introduces another factor, since it is essential that emergency power to other functions in the accommodation should not be jeopardised. In short, the ventilation system needs to be thought through to minimise the chance either of its being ineffective or defeated or of its actually making things worse.

19.172   There is, however, one specific measure which I am satisfied I should support. The air intakes of the ventilation system should be provided with hydrocarbon gas and smoke detectors and on alarm the ventilation and dampers should shut down. I note that the draft fourth edition of the guidance notes to the Construction and Survey Regulations contains provisions on this matter.

19.173   In due course assessment of smoke ingress into the accommodation will be part of the Safety Case. Meanwhile, such an assessment is one of the measures which the regulatory body should ask operators to undertake forthwith.

19.174   Escape routes also are a proper topic for treatment in the regulations on construction, which should set the goal that the escape routes should be passable. The regulations should allow an approach integrated as between the twin threats of fire and smoke, as between the different options for protecting the escape route itself, and as between protection of the route and use of personnel protective equipment. Embarkation points should be treated in a similar way.

19.175   There is one set of specific requirements which it is appropriate for the regulations on construction to include. This is that escape routes be provided with adequate and reliable emergency lighting and with photoluminescent direction signs which are not dependent on survival of power supplies.

Emergency centres and emergency systems

Emergency centres

19.176   The next topic which I wish to consider in this chapter is emergency centres and emergency systems. As the emergency on Piper developed there were no facilities in the ERQ to assess or exercise control over it or to communicate with the outside world. The Control Room was knocked out and was in any case outside the ERQ as was the Radio Room, which was abandoned at an early stage. There was no means of obtaining information from, or of determining the status or action of, any of the emergency systems such as the F&G detection, ESD or deluge systems. In an attempt to discover what was happening people opened doors, which led to further ingress of smoke into the accommodation.

19.177   The need for facilities within the TSR which would allow the occupants to assess the situation outside and to exert some control over it was one of the points made by Mr Brandie. He instanced the need for communications and controls, including fire-fighting facilities. He envisaged that the TSR would contain emergency power generation and fire pumps. Controls were most readily available if the control room was located within the TSR. This was his preferred solution for new platforms, though he recognised that on some existing platforms the control room was outside. Mr Booth stated that in his company the control room was always in the TSR and that this was the practice on modern platforms. Both witnesses also envisaged the TSR as containing the radio room. The general trend described by these and other witnesses was towards locating the radio room in the accommodation and to locating the control room either in it or readily accessible from it.

Observations on emergency centres

19.178   I believe there is a clear need for there to be available within the TSR certain minimum facilities for controlling an emergency. There should be means of internal and external communication, of obtaining information on what is going on outside, and of exercising at least some degree of control over it. In general terms, what I have in mind here is information on key process, pipeline and fire system variables and on the operation of the ESD system together with certain key controls on these systems.

19.179   Most of these minimum facilities already exist in the control room and the radio room. It is logical therefore to locate the control room and the radio room in the TSR. This ensures that the facilities are accessible and protected.

19.180   Where on an existing installation the control room, the radio room, or both, are outside the TSR, the minimum facilities need to be made available in the TSR. This requires that there should be created within the TSR an emergency control centre, an emergency radio room, or both, as the case may be, which contain the necessary minimum facilities. The fuller facilities in a control room or radio room outside the TSR may still be valuable and any such rooms should be protected and should have secure means of communication of information with their opposite numbers in the TSR.

19.181   It is not intended that either of these rooms should act as the emergency command centre, which should also be within the TSR but in a separate room.

19.182   I make no proposals on the precise nature of the minimum facilities which should be made available within the TSR. A radio room and a control room which is designed to allow control as well as monitoring are likely to contain most, though perhaps not all, of the facilities required. The provision of these minimum facilities within the TSR should be part of the Safety Case and their selection should be specified by the operator.

Emergency systems

19.183   Turning to the emergency systems, whilst there is some uncertainty as to the precise extent and cause of damage to individual emergency systems on Piper, the general picture is clear. Both the main and emergency power supplies were knocked out, and possibly some of the UPS. Battery power supplies dedicated to individual equipment mainly performed well. The main means of communication to the generality of people on the platform, the GA/PA system, may have been disabled, though this is uncertain. In any event it was not used and the other means of internal communication such as telephones and hand-held radios were no substitute. One of the main means of external communication, the tropospheric link, lost its power supplies, but in this case the existence of alternative links allowed communications to be maintained as long as the radio room functioned.

19.184   Evidence on electrical power supply systems was given by Mr J Day, Head of Electrical Engineering of Shell Expro, with special reference to maintenance of the integrity of emergency power supplies. He took the view that in an accident it was not unlikely that the main generators might be lost and that effort should be concentrated on ensuring that the emergency supplies remained available. The general approach was to classify equipment and services by priority and to match the integrity of the emergency supplies to those priorities. He described modern developments in high integrity generation, implemented on Kittiwake, including fire pumps driven solely by electricity and without diesel back-up, and in recombination cells for battery power supplies.

19.185   Mr Day was asked about protection of electrical power supplies against the effects of an explosion and particularly against severe platform vibration, or shock. He said he did not know of any case where his company was designing to protect against an explosion. He was unaware of any case where shock, whether from explosion, vessel bumps or dropped objects, had caused loss of electrical systems, other than what may have happened on Piper. Mr Nordgard confirmed that vulnerability of the electrical supplies to platform vibration was not something to which particular attention was paid.

19.186   Mr Day referred to the statutory requirement to provide emergency power to a minimum of 24 hours. He was asked whether any relaxation of the time period would open up the possibility of alternative means which would give a more reliable emergency supply, but he did not believe it would.

19.187   A number of witnesses in Part 1 suggested that it would be helpful if there was a greater degree of uniformity in the alarm systems for emergencies. This would be of particular assistance to contractors' personnel, moving as they do from installation to installation. The status light systems used on some platforms were also advocated.

Observations on emergency systems

19.188   It is clear that great efforts are made to maintain power supplies in accident conditions, but it is also clear that if the accident is a severe one, even the emergency supplies may be vulnerable to effects caused by the accident. This vulnerability is shared by the emergency systems generally and I have already referred to the vulnerability of ESVs and of fire protection systems. I am concerned that all emergency systems should possess in a high degree the ability to survive severe accident conditions. The emergency systems which I have in mind include the emergency power supplies and systems, the ESD system and the communications systems and the severe accident conditions to which I refer are primarily fire, explosion and strong vibration.

19.189   The regulations and guidance should promote an approach to the design of emergency systems which ensures that as far as is reasonably practicable the systems are able to survive severe accident conditions. The ability of these systems to survive severe accident conditions should also be a feature of the Safety Case.

19.190   Work needs to be done to determine the vulnerability of emergency systems to severe accident conditions and to enhance their ability to survive such conditions.

19.191   In due course assessment of the ability of the emergency systems to survive severe accident conditions will be part of the Safety Case. Meanwhile, such an assessment is one of the measures which the regulatory body should ask operators to undertake forthwith.

334

19.192   I believe there is merit in the status light systems which are installed on some platforms and would wish to see them promoted.

19.193   I note that status light systems have the characteristic that they are, or can be designed to be, fail safe. That is to say, they can still convey their essential message even on loss of power. This is a feature which can be crucial in accident conditions and which would seem to have application to other aspects of platform communications. The regulations should promote this general concept.

19.194   I accept that a greater degree of uniformity in the status light systems and the alarm systems for emergencies on installations would be helpful. I can see no argument for not trying to achieve standardisation.

Pipeline emergency procedures

19.195   The disaster on Piper revealed deficiencies in command in emergencies. It also revealed deficiences in the emergency procedures for the other platforms connected to it by pipeline. I will consider in Chapter 20 all matters related to command and procedures for emergencies on the platform affected by an accident and confine myself here to the pipeline emergency procedures. It was clear from Part 1 of the Inquiry that the emergency response of platforms connected by pipeline to a platform affected by an accident is a problem area. No further evidence on this topic was led in Part 2.

Observations on pipeline emergency procedures

19.196   The quality of pipeline emergency procedures needs to be improved. There should be more co-operation between operators in a field in the formulation of arrangements and the writing of manuals. There should also be more involvement in these activities by the personnel most directly affected, those on the installations, to ensure that the information contained is correct and that the procedures proposed are the most practical and effective. The procedures should be reviewed regularly and the manuals updated in a co-ordinated manner.

19.197   The pipeline emergency procedures for the installation should define the conditions which constitute reason to believe that there has been an incident on another installation connected to the first by hydrocarbon pipeline and the conditions for shutdown of the first installation. The overriding aim should be to ensure that the situation on the affected installation is not exacerbated. In general, shutdown should be the default action and should be effected at once unless it can be positively and reliably confirmed that the incident on the other installation is minor. The shutdown procedures should be reviewed regularly and the manuals updated.

Observations on mitigation of incidents

19.198   I said at the beginning of this chapter that I was conscious of the calls which have been made for there to be requirements for various types of hardware. I have in fact made very few such proposals. In a limited number of cases I have taken the view that it is inconceivable that a particular measure should not be adopted. The requirements in the recent regulations on ESVs are a case where I do consider that this is so. In general, however, I have found that the matter is one which should be decided on the basis of the resolution of the often conflicting factors which design typically involves and of what is reasonably practicable. For each particular issue I have explained this as I went along. Here I wish to make a further point. The decisions on the various hardware proposals cannot be viewed in isolation. This is another argument for dealing with these matters in the Safety Case.

19.199   Finally, it is convenient to deal here with a general point concerning acceptance standards, particularly those for protective systems such as the fire-water deluge system and for emergency systems such as the emergency power supplies. I am proposing reliance on goal-setting regulations and on the Safety Case and eschewing prescription of hardware. Such an approach therefore depends heavily on the acceptance standards for achievement of the goals. In general I propose that these should be set by the operator. I envisage that it is around these standards that much of the dialogue between the operator and the regulatory body will centre.

Table 19.1 - Study of installation configurations: average annual fatalities in high fatality accidents

| Event | Fatalities per 1000 installation years Case | | | | | |
	A	B	C	D	E	F
Blowouts	7.9	7.5	6.7	4.9	1.3	4.8
Riser failures	30.2	6.7	14.4	13.2	2.3	8.8
Process leaks	40.4	22.0	15.8	18.7	9.2	10.3
Collisions	4.6	5.0	11.3	55.3	1.9	5.8
Structural events	82.7	9.5	40.0	50.6	3.5	23.0
Non-process fires	0	0	0	0	0	0
Helicopter accidents	0	0	0	101.2	0	0
Total	165.8	50.7	88.2	243.9	18.2	52.7

Notes:

(a) Case E has a production rate one quarter that of the other cases. The complements are 200 persons on cases A and B, and extra 20 marine crew on the flotels in cases C and D, and an extra 5 maintenance crew on the quarters platform in case F; case E has a complement of 60 persons.

(b) Case D includes fatalities from in-field helicopter accidents, regardless of the number of fatalities. Fatalities during crew changes are not included for any of the cases, since they would be the same for all.

Chapter 20

Evacuation, Escape and Rescue

Introduction

20.1   In Chapters 8 and 9 I described how the personnel on board Piper responded to the emergency on the night of the disaster, escaped from the platform and were rescued from the sea. In this chapter I comment on the requirements and regulations for safe evacuation, escape and rescue and review the arrangements and facilities used for them in North Sea conditions. Finally, I discuss the requirements for effective command and control in offshore emergencies.

Evacuation, escape and rescue: definition

20.2   To avoid confusion or doubt, the scope of the terms of evacuation, escape and rescue, as used throughout this chapter, are defined below:

Evacuation refers to the planned method of leaving the installation without directly entering the sea. Successful evacuation results in those on board the installation being transferred to an onshore location or to a safe offshore location or vessel.

Escape refers to the process of leaving an offshore installation in the event of part or all of the evacuation system failing, whereby personnel on board make their way into the sea by various means or by jumping.

Rescue refers to the process by which escapees and man overboard (MOB) casualties are retrieved to a safe place where medical assistance is available.

History of evacuation, escape and rescue

20.3   Dr J Side of the Institute of Offshore Engineering, Heriot-Watt University, described 4 major offshore incidents, all occurring outside the UKCS, that have themselves been the subject of Government inquiries. All the incidents occurred to mobile offshore structures. In March 1980 the semi-submersible accommodation vessel, the Alexander Kielland, sank in the Ekofisk field off Norway due to a structural failure. The considerable heel of the structure when it capsized made the launching of survival craft extremely difficult but attempts were made to launch 5 of the 7 craft on board. 3 were crushed against the side of the rig and destroyed. One which had not been launched but had been entered, came to the surface inverted after the rig had capsized and was eventually righted. The fifth craft, with 26 men on board, released its hooks with considerable difficulty. Of the 212 men on board, 123 lost their lives. In February 1982 the semi-submersible drilling rig, the Ocean Ranger, capsized and sank in a very bad winter storm off Newfoundland. Warnings that the weather would deteriorate had been broadcast 24 hours before the disaster occurred and drilling operations on the rig had been discontinued some 12 hours before the incident. However a mayday requesting helicopter evacuation was not sent until 2 hours before the rig capsized. At least one survival craft was launched but although the standby vessel made contact with it no one could be saved. The entire crew of 84 persons was lost. In October 1983 the drillship Glomar Java Sea capsized and sank in the South China Sea during a severe typhoon, with the loss of its entire crew of 81 persons. The investigation found that the failure to evacuate at least non-essential personnel from the drillship, after it had been given nearly 3 days' warning of the typhoon, was a contributory factor leading to the loss of life. Examination of the wreckage suggested that attempts had been made to launch one lifeboat; another had torn free as the vessel sank. Neither lifeboat was ever recovered. At least 36 of the 81 crew were trapped in the drillship as it sank. In October 1985, in the Gulf of Mexico, the jack-up drilling rig Penrod 61 collapsed during a hurricane. Soil failures beneath one of its 3 legs apparently caused the failure. The standby vessel, a normal crew boat, which was unable to operate safely in very severe weather, had to leave the area to seek refuge from the worsening weather about 9 hours before the rig collapsed. The crew of 41 escaped to the sea in 2 survival capsules and an inflatable raft; one man jumped into the sea with a life-jacket. Only one life was lost when one of the survival capsules subsequently capsized. In all the last 3 incidents weather conditions at the time of the accident had deteriorated to the point when helicopter evacuation was impossible.

20.4   Dr Side also described 18 precautionary evacuations in the UKCS between 1975 and 1987, following incidents that could have led to a major emergency on an offshore installation. Common to all of these was the immediate requirement for an urgent, unscheduled demanning of personnel. Except for 3 transfers by personnel basket to a vessel, and one in which the means of evacuation remains unknown, the rest were by helicopter. Survival craft were not used. In August 1988 the first full emergency evacuation using survival craft occurred when the crew of the semi-submersible drilling rig Ocean Odyssey had to abandon the rig after a blowout and fire. One man died but the actual evacuation, in good weather, went smoothly. Previously in the NCS there were 2 recorded cases of survival craft being used, apart from that of the Alexander Kielland described in para 20.3. In November 1975 an explosion on the Ekofisk A platform led to an evacuation. The platform shut down automatically and the situation was brought under control but due to the failure of its launching mechanism a survival craft was dropped from the deck level and 3 men were killed. In March 1977 another Ekofisk platform was evacuated by 112 men using 3 survival craft following a blowout. The evacuation was orderly and disciplined; and the weather and sea state were remarkably good at the time.

20.5   In none of the 18 precautionary evacuations in the UKCS did the initial incident develop into a major emergency but the disasters which resulted from delayed reactions to weather warnings demonstrate how severe the resulting penalty can be. They provide the lesson that where there is doubt as to the implications of an incident it is better to achieve certain safety by a precautionary shutdown and evacuation than to risk lives by postponing the decision. A precautionary evacuation will normally be by helicopter which is the preferred and widely available method for such circumstances. As I show later in para 20.14, if a major incident has already developed such a method is unlikely to be available; and evacuation will depend on the uncertainties inherent in the use of survival craft. The decision to evacuate an installation as a precautionary measure is dependent on the perception of the OIM. I will discuss command and control in emergencies later in this chapter (see paras 20.56 *et seq*).

General approach

Requirements and regulations for evacuation, escape and rescue

20.6   The objective of ensuring safe evacuation, escape and rescue on offshore installations was summarised for the Inquiry by Mr Petrie, Director of the Safety Directorate of the DEn, in saying that "Offshore installations should be designed, equipped and organised so as to provide means of safe evacuation of all personnel on the installation in the widest practicable range of circumstances" and that "the means of evacuation should be available for immediate use." The requirements for safe evacuation, escape and rescue are covered by Reg 10 of the Emergency Procedures Regulations and the code of practice for the assessment of the suitability of standby vessels attending offshore installations; and by the Life-Saving Appliances Regulations. As a result of lessons learnt from the Piper disaster the DEn has proposed amendments to the last-mentioned regulations in a statement of intent issued in August 1989.

20.7   Mr Petrie explained that where a bridge link to an adjacent installation was available, this was the preferred means of evacuation. Leaving aside a bridge link, helicopters represented the safest means of evacuation provided that there was sufficient time and no conditions adverse to evacuation by this means existed, such as fire, smoke and emission of combustible gas. The circumstances required for helicopter evacuation might not prevail when an emergency had developed; and installations must be provided with another primary evacuation system, wholly controlled from within the installation, not dependent on any external intervention, and capable of securing safe and full evacuation of all personnel in as wide a range of emergencies as practicable. This evacuation system was based on totally enclosed motor propelled survival craft (TEMPSC). He said that if for any reason these primary systems were partially or wholly unavailable there should also be provided means of descent to the sea and the means of rescuing people from the sea, which should be effective in the widest possible range of weather conditions. Appropriate personal survival equipment should be provided for all the personnel on board. I will deal with all these requirements individually later in this chapter.

Evacuation, escape and rescue and the Safety Case

20.8   In discussing the mitigation of incidents in Chapter 19 I expressed the view that the integrity of the temporary safe refuge (TSR) should be a central feature of the Safety Case (see para 19.109). It is plainly appropriate that the process of formal safety assessment should cover all aspects of the protection of personnel in the event of an emergency and therefore should cover the process of evacuation, escape and rescue. This should accordingly form part of the Safety Case. I would note that Mr Ognedal, Head of the Safety Division of NPD, emphasised that under the Norwegian regulatory regime the operator is required to have a thought-through evacuation and escape philosophy which formed part of the framework of the whole system for the safety of personnel on board.

Evacuation, escape and rescue and the regulations

20.9   Evidence on the regulatory requirements for safe and full evacuation, escape and rescue was heard from UKOOA, the DEn, the NPD and Statoil. UKOOA urged that due to the diversity of installations offshore, the present prescriptive regulations should be replaced by goal-setting regulations which would allow operators more flexibility and would not stifle innovation. Mr Ognedal stated that future Norwegian regulations on evacuation, escape and rescue would be mainly goal-oriented; the emphasis would be on the licensee identifying the best evacuation systems for the installation in question, through purposeful and systematic analysis. Prescriptive regulations would, however, be retained in a few selected areas. The DEn accepted, in general, the replacing of specific requirements by general principles but would also wish to retain specific requirements in defining acceptable standards in certain well-defined cases. I fully support the acceptance of goal-setting regulations in this area. In particular I see an immediate need for regulations under which operators are required to submit to the regulatory body an analysis of the facilities and other arrangements which would be available for evacuation, escape and rescue in the event of an emergency. The analysis would cover the formal command structure, helicopters, TEMPSC, life rafts and other means of escape to the sea, standby and other vessels, fast rescue craft and personal survival and escape equipment. This analysis would also form a part of the Safety Case in its demonstration that adequate provision had been made for ensuring safe and full evacuation, escape and rescue. Operators which have not already done so should be asked to undertake such an analysis without waiting for legislation. While I fully support the acceptance of goal-setting regulations I also take the view that there are certain basic points on which certain minimum standards should be laid down by legislation. Examples of these are given later in this chapter.

Evacuation by helicopter

20.10   Helicopters are the normal means of transport for personnel to and from offshore installations. Everyone working offshore accepts the use and discipline of helicopter travel and would automatically look to helicopters as the prime means of evacuating them from an installation. There are no specific regulatory provisions compelling the use of helicopters although there are provisions in the Construction and Survey Regulations for ancillary matters such as landing arrangements and fire-fighting facilities, and the accessibility of the helicopter deck from the accommodation.

Helicopter availability

20.11   Evidence on the performance and availability of helicopters on the North Sea was given by Captain Ginn, Head of Air Transport for British Gas. Availability is high offshore. Commercial traffic in normal day-time working hours between 07.00 and 18.00 hours on week-days would enable between 100-300 helicopter seats to be located in any one of the 4 sector areas in the northern and central North Sea and thus available to an emergency in less than 30 minutes. All helicopter operators require their crews to make "operations normal" calls into base every 20 minutes or so; and one helicopter company maintains a "flight following" system whereby the identification, position and full status of each machine in the air is entered routinely into a computer at base. If an offshore emergency were to occur, the identity and location of the nearest helicopter and time required to reach the emergency site can be obtained immediately. The shore-based national air traffic service radios do not reach the more distant areas of the North Sea; in the East Shetland basin the Viking Approach system on Cormorant Alpha supplements the shore-based service. Some companies use, or are planning to use, flight information liaison officers (FILOs) on their installations, with whom the pilots of helicopters make contact so that the installation has a constantly up-dated record of helicopter availability in its area. At the time of the disaster Piper was the only installation in the North Sea utilising FILOs. It appears to me that a North Sea-wide flight following system based onshore and operated as a service to all offshore installations would be more efficient than the duplication inherent in individual installation systems but I appreciate that operators may feel more secure with their own systems.

20.12   In addition to land-based helicopters, offshore helicopters are based for logistical purposes on 5 UK and 2 Norwegian North Sea installations. These offshore-based helicopters are smaller (4-13 seat capacity), are crewed on a 24 hour basis, and except for 2 in the UK sector, are equipped for search and rescue. The cost of basing helicopters offshore is very high, amounting to over £2m per year per helicopter for the charter of the helicopter and the provision of crew accommodation offshore. This does not include the capital costs of modifying the facilities where there may be severe structural constraints, particularly on existing installations. Search and rescue helicopters are also available from a number of onshore military establishments as well as from 2 civil helicopter bases maintained by the DoT, at Sumburgh and Stornoway. The declared response times (call-out to take-off), often bettered in practice, are 15 minutes by day and 45 minutes at night. Commercial helicopters are also available after normal working hours from the Aberdeen, Sumburgh and Unst bases and would be despatched to an emergency within 10-30 minutes of call out. In an emergency it is most likely that the evacuation of an installation by helicopter would be initiated by offshore-based logistics helicopters and/or en route commercial helicopters, depending on the time of day. This would generally be in less than an hour. Out of normal working hours (Mondays to Fridays, 07.00-18.00 hours) the first helicopters to arrive would normally be the offshore-based helicopters, in less than one hour, followed by the onshore-based machines. The most rapid evacuation method would be to shuttle personnel on board to nearby installations; or it may be possible to fly them directly to shore. Shortage of helicopters is unlikely to be a problem; access to the helideck becomes the limiting factor when the number of helicopters in a shuttle reaches say 4 or 5. The cycle of a helicopter approaching, landing, and boarding up to 19 evacuees (in the case of the Super Puma) and then taking off is unlikely to average less than 5 minutes.

20.13   The evacuation of 6 typical installation types by helicopter was assessed by UKOOA who found that the maximum time to evacuate (by shuttling to nearby installations or to shore) would be in the order of 2 hours and 45 minutes from call-out for the worst case (an isolated drilling unit at night). Two actual precautionary evacuations, both in late 1989 (North West Hutton and Penrod 92), confirmed that the findings of the study were realistic. Both were in daytime: one was totally evacuated 70 minutes after call-out (110 men), the other in 40 minutes (70 men). The first helicopters arrived in 23 and 5 minutes respectively.

Helicopter evacuation limitations

20.14   Evacuation by helicopter may be limited by adverse weather conditions such as very high winds, low fog or icing, although according to statistics these limiting conditions seldom apply. Helicopters are allowed by their operators to land and take off in winds of up to 60 knots (Beaufort Force 10) although they must keep their rotors moving. In the case of emergency evacuation, however, pilots would be expected to find a clear way to the installation and be forced to abandon their attempts only when the limits on the air-worthiness of the aircraft were reached. Statoil estimates that in the Norwegian sector the availability of helicopters is 98.7% when the platform is evacuated as a safety precaution (see para 20.5). Major incidents, however, normally result in large amounts of fire, smoke, and/or flammable or combustible gas being generated. This would prevent a helicopter approaching or landing on the installation. There may also have been a major structural failure which would prevent a landing. Safety studies done on integrated production platforms in the Norwegian sector show that when such emergencies have developed evacuation by helicopter would be impossible in about 95% of such cases. If these conditions prevail, an alternative primary evacuation system using TEMPSC will be necessary. In an extreme case, resort to direct entry into the sea will be the only remaining means of reaching safety.

20.15   Helicopters remain the preferred method of evacuating an offshore installation as a safety precaution measure and in the limited instances in which they can be used in a developed emergency. They are quickly available on the North Sea, safe in all but the most adverse weather conditions and offshore personnel are accustomed to their use and discipline. The people evacuated are transferred directly to places of safety and are not left still at the mercy of weather conditions, as in evacuation by survival craft.

Evacuation by survival craft (TEMPSC)

20.16   Reg 5 of the Life-Saving Appliances Regulations requires that every normally manned offshore installation shall have TEMPSC of sufficient capacity in aggregate to accommodate 150% of the entire platform complement. In its statement of intent the DEn proposed increasing the required capacity of TEMPSC to accommodate twice the number (200%) of persons on the installation, "to enhance the safety of personnel on board an offshore installation". This proposal was evidently made in reaction to the disaster. I have 2 main difficulties with this proposal. Firstly, the non-availability of TEMPSC at the time of the disaster was not related to their number as such but to their location and distribution. Secondly, the proposal takes no account of the features of particular installations. On installations which have certain complexities and configurations it may be desirable to provide a wider distribution of TEMPSC to improve the range of circumstances in which a safe and full evacuation is likely to be achieved. I would favour the retention of the existing requirement to accommodate 150% of the entire platform complement. However I consider that it should be required that the TEMPSC provided should include TEMPSC which are readily accessible from the TSR and which have in the aggregate sufficient capacity to accommodate on board the number of persons on the installation. The exact number, location of any TEMPSC which may be required over and above these minima should be determined in the light of the Safety Case. It may, for example, be shown that additional TEMPSC should be provided near places where personnel may congregate or be trapped in an eventuality.

Davit-launched TEMPSC

20.17   The original design of offshore survival craft was a standard ship's lifeboat, with a full canopy and water deluge system added. A UKOOA witness, Mr I Wallace of Conoco (UK) Ltd, commented that the design of survival craft was and still is constrained by the International Maritime Organisation (IMO) standards; they were still seen as ship's lifeboats and had not been adapted to the requirements of the offshore oil industry. He said that a number of improvements more suited to the industry's specific needs had been put forward but the lifeboat manufacturers continued to design to IMO standards because the bulk of their business was concerned with ships.

Problems clearing the platform legs

20.18   A critical problem with a davit-launched TEMPSC is ensuring that it gets away from the vicinity of the installation after launching into the sea in severe weather conditions, when it would be in danger of being swept under the installation and destroyed. Also wind-induced motion could cause the TEMPSC to contact the installation during descent, if the overhang is less than 7m. The minimum weight of a TEMPSC is determined by this latter problem. The DEn's statement of intent acknowledged these concerns. Since 1985 a joint steering committee, with representation from Government departments, industry and contractors has been considering enhanced launching techniques for survival craft. Two passive devices, one using a hinged boom projecting from the side of the installation to rotate the TEMPSC outwards, called PROD, and the other using an air-launched sea anchor, are being tested. So also is a powered dolphin similar to a torpedo. The DEn and the industry have supported full-scale trials of the PROD concept and it is being developed into a commercial product. Joint work by an oil company and a contractor has led to the testing of another concept, in which an anchored buoy is used to direct the TEMPSC away from the installation, called TOES. No enhanced launching system has yet been proposed for an installation in the North Sea.

Problems with davit disconnection

20.19   All TEMPSC are boat-shaped and thus launched on 2 fall wires (bow and stern). It is essential that the hooks on both wires should release the TEMPSC simultaneously. Two basic release systems have been used, the "off-load" system which will only allow the hooks to release the TEMPSC when it is afloat, and the "on-load" system which allows the simultaneous release of both hooks. The on-load system must be used in conjunction with hydrostatic interlock, which allows hook release only when the hull reaches the water; premature release could be disastrous, as in the attempted evacuation of the Ekofisk A platform in 1975 (see para 20.4) when a loaded TEMPSC was accidentally dropped into the sea from deck level. The off-load system has the major disadvantage that in severe weather conditions it is difficult to achieve a condition whereby both hooks are off-load. This has now been resolved by IMO resolutions which specify that survival craft should be equipped with on-load release gear. However the difficulties experienced in North Sea operations with the premature release of davit fall wires contributes to the prejudice felt by oil workers offshore who, in the main, are inexperienced in marine matters.

Free-fall lifeboats

20.20   A new type of survival craft, the free-fall lifeboat, has been introduced into the Norwegian sector. The free-fall lifeboat is a TEMPSC designed to be used with a launching system which releases the TEMPSC at the point of embarkation on the platform (up to 30m above sea level) and allows it to fall, entering the water with a high momentum which together with the specially shaped hull will propel the craft away from the platform. The particular perceived advantage of the free-fall lifeboat over the davit-launched lifeboat is its ability to clear the installation in severe weather conditions.

20.21   Development of the free-fall lifeboat started in 1973 but this was accelerated after the Ekofisk Bravo blowout in 1979 when a large research project was funded by the NPD. A total of some 40 free-fall lifeboats had been or were being installed on Norwegian installations at the time of the Inquiry. Free-fall lifeboats are now considered in the Norwegian regime to be the established evacuation system. Conventional davit-launched lifeboats are accepted but their availability must be shown to match that of free-fall boats. The Norwegian witnesses explained that all factors, mechanical as well as weather, are taken into account in this. Free-fall boats have been model-tested in severe sea state conditions and their overall availability is estimated at 99%. The remaining 1% is attributable to technical and human considerations.

20.22   Norwegian offshore personnel on installations on which free-fall lifeboats are installed must undergo special training which involves at least one fall in a free-fall lifeboat at one of 3 training centres. As an example of their reliability, it was said that one free-fall lifeboat at a training centre had been dropped some 1200 times without an accident of any kind. The occupants of free-fall lifeboats lie on their backs on special contoured seats and are restrained by body and head straps. In the sea this position is not comfortable and can accentuate sea sickness but the benefits are considered to outweigh the discomfort. The standard capacity for free-fall boats installed on large integrated platforms is 74 people.

20.23   Free-fall lifeboats, apart from having the advantage over conventional lifeboats of safely clearing the installation when launched in severe weather conditions, have no davit hooks to be unlatched. The total time from the embarkation decision to water entry is about the same as for davit-launched boats; more time is needed to seat and strap in the occupants but time to the sea is much less. They are, however, much heavier and more expensive than davit-launched boats and retrofitting can be a major problem. They have not as yet been used in actual emergencies in severe weather but extensive model testing has made the Norwegian authorities confident of their high rated availabilities.

20.24   No legislation directly regarding free-fall lifeboats so far exists in the UK. Reg 5 of the Life-Saving Appliances Regulations entails that the means of launching lifeboats should be by lowering, but the DEn's statement of intent proposes that this be amended to allow the launching of TEMPSC by any safe launching system of a type which has been tested and is acceptable to the DoT. It was stated that this would permit consideration of free-fall TEMPSC without the need for exemption; and that exemptions had already been granted for free-fall TEMPSC on a small number of mobile installations in the UKCS. The DoT had advised the DEn that free-fall lifeboats are a viable method of evacuation. I see no reason why free-fall lifeboats should not be permitted in the UKCS, as they are in the NCS. The necessary amendment to the regulations should be made forthwith. There should be no statutory barrier to the use of free-fall lifeboats. It would still remain for the operator to justify its choice of TEMPSC as being appropriate in the particular conditions of its installation. Where davit-launched TEMPSC are proposed to be used they should be oriented so as to point away from the installation.

Escape to the sea

20.25   Reg 8 of the Life-Saving Appliances Regulations requires that alternative means of evacuation to the sea have to be provided, so that the fullest practicable evacuation can still be secured. Below I describe various means in common use and comment on some new means.

Life rafts

20.26   Life rafts are not considered by the DEn to be an acceptable substitute for the required provision of TEMPSC. They do, however, usefully complement TEMPSC. The DEn statement of intent proposes to amend Reg 5 of the Life-Saving Appliances Regulations to require that, in addition to TEMPSC, every normally manned installation should be provided with life rafts having sufficient capacity in aggregate to accommodate all the people on board. I have already recommended (see para 20.16) that the total TEMPSC capacity on an offshore installation should remain at 150% of POB and be more if required by the Safety Case for the installation. The proposal to require 100% POB life raft capacity seems to me to be a reasonable requirement, as life rafts can complement TEMPSC in the event of emergencies such as a sudden structural failure, keeling over or sinking, where access to TEMPSC is prevented or their use is impracticable. The location of life rafts would be a subject for the Safety Case for the installation. They should, however, be installed in close proximity to mechanical means of escape to the sea such as ladders, ropes and escape chutes.

20.27   After an examination of the life rafts recovered from Piper and in the light of evidence that survivors were unable to deploy them, the DEn issued a safety notice (9/89) emphasising the existing legal requirement that life raft launch procedures should be included in musters and drills and clarifying the position with regard to the length of painter rope (cf paras 8.29-33). Painters will be shortened to the minimum practical length to ensure successful deployment; and the length of painter to be provided at each life raft launch point is to be agreed with the DoT.

Ladders and stairs

20.28   In their statement of intent the DEn recommend the installation of permanent ladders or stairs to the sea at the corners of the installation. Mr Wallace noted that it had been the practice to have 2 or more constructed ladders or stairways leading to the sea but that these suffered from storm damage and the effect of waves. In view of the difficulties experienced by personnel in getting safely into the sea at the time of the disaster, I support the DEn's recommendations. These ladders or stairways could be extendable to allow for the effect of waves provided this is acceptable in the Safety Case.

Ropes and rope devices

20.29   Knotted ropes, rope ladders and scrambling nets are very basic means of descent to the sea and have long been used on offshore installations. However only knotted ropes would appear to be practical as they can be easily and economically stowed at all life raft installations. Scrambling nets and rope ladders are awkward and difficult to use. In the disaster approximately half of those who escaped to the sea did so by using ropes or small diameter hoses. The others jumped where haste was imperative and/or ropes were not available. Knotted ropes are, however, a primitive means of escape and physically very demanding. They are particularly difficult to use if the user is wearing a life-jacket and/or a survival suit. But they are almost foolproof, they offer a continuous line of escape and place a person outside the confines of the installation structure. They should be seen as a last means of escape to the sea. A new individual self-rescue device, based on abseil technology and much less demanding physically, the Donut system, has been accepted for use by both the DEn and the NPD. At least one UK operator is issuing it generally to offshore staff. Some training is required and it cannot be re-used. I recommend that such equipment should be specified in the regulations for escape equipment. Another rope-based system is the Surescue Descender, an escape line down which people can descend on light-weight supports in a controlled manner. It accommodates one person at a time and would not seem to offer rapid evacuation for a large number of people. The Surescue is, however, acceptable to the DoT and one such system is installed on an accommodation barge in the North Sea.

Escape chutes and collapsible stairs

20.30   A number of these devices were described at the Inquiry. One, the Skyscape, a collapsible tubular net which allows controlled free-fall descent to the sea, has been accepted for use by the NPD and has been installed on a Norwegian installation. It requires an overhang and may not be suitable for all installations. The GOTECH escape stair is at the conceptual and scale model stage and much work remains to be done on it. These chutes or stairs have potential advantages over fixed ladders or stairs in that they would offer direct escape to a large number of personnel; they are positioned away from the platform, hanging from an overhang or cantilever; and they appear to avoid the difficulties inherent in the use of knotted ropes. Other chute-type devices were briefly described, none apparently yet available for full scale testing. One operator, Shell Expro, considered chute escape devices in the design of their latest platform but decided that due to certain disadvantages they were not yet appropriate.

Other devices

20.31 Another operating company, Mobil, is developing 2 dry transfer devices with different manufacturers. One (GEMVAC) is similar to the system used by navies for the transfer of materials and men at sea. The other (ODELE) involves lowering an inflatable life raft with men on board on to a specially constructed SBV. Both are in the development stage. Other devices, all at early conceptual stages, include a sea haven and a self-launching accommodation module.

Personal survival equipment

Life-jackets and survival suits

20.32 Life-jackets for all personnel offshore are required by regulation. The DEn proposes in its statement of intent to amend the regulations to require that offshore installations are provided with at least twice as many jackets as the number of people on the installation. The provision of survival or immersion suits is desirable but not yet mandatory. It was stated in evidence that the DEn was seeking through further work with the DoT and other organisations to overcome certain practical difficulties about the wearing of life-jackets with survival suits which had been identified. When this work was complete it was expected that provision of survival or immersion suits on all offshore installations would be made mandatory. Mr Wallace described an integrated survival suit and life-jacket which he said was in use by his company, Conoco. In view of the comments by survivors on the difficulties of making escape while wearing conventional life-jackets and immersion suits I recommend that this planned work should be carried out with despatch.

20.33 The use of a standard orange colour for life-jackets and survival suits was criticised by many engaged in the rescue of men from the sea; other objects and equipment such as life rafts are of the same colour. Attention must be given to using a separate and distinct colour for easy and rapid identification of survivors in the water, particularly in the dark.

20.34 Comment was also made on the problems of locating survivors in the sea. Whistles and lights are supplied on most life-jackets but may not be effective in adverse weather conditions. The use of radio transmitters or detectors should be considered.

Smoke-hoods

20.35 The ability to move through a smoke cloud can be of vital importance in an evacuation or escape. The escapee will have to move from his location at the time of the incident to the TSR or directly to an embarkation point. Very large quantities of dense, and possibly toxic, smoke are likely to be generated from a fire on a hydrocarbon producing installation. Evidence heard on smoke-hoods generated some discussion on the period for which simple filter-type hoods would be effective before the breathed air is over-saturated with carbon dioxide. Expert evidence was to the effect that filter-type smoke-hoods could provide temporary respiratory protection against smoke and toxic gases. It was suggested that this could be for some 10-15 minutes after which the concentration of carbon dioxide could seriously debilitate the wearer. Oxygen donating hoods are complex; special training for use is required. They are also bulky and expensive but can be very useful for exploratory investigation of an accident by specially trained men. Simple, light, filter-type hoods should be issued as part of personal survival kits to be kept by all on board offshore installations (para 20.36) but training should emphasise that they provide protection only for a limited period to facilitate evacuation or escape. On-going research sponsored by the industry on the development of improved smoke-hoods should be expedited.

Survival packs

20.36   Some operating companies, among them Occidental, issue survival packs to those going offshore. Generally these packs contain a life-jacket, a survival or immersion suit, a torch and fireproof gloves. Such equipment packs should become standard issue offshore. The packs can be kept in the individual's living quarters and/or at work sites and regularly examined. A type of smoke-hood, as described above, should be added to the kit, as well as any other simple and personal survival aids that may become available. I recommend that survival packs containing at least a personal survival suit, a life-jacket and a smoke-hood should be issued to everyone on board an installation and that these are normally retained in their accommodation. Other survival packs for at least half the POB should be stored in containers placed at locations on the installation subject to what is shown necessary by the Safety Case.

Rescue from the sea

Standby vessel (SBV) legislation and code of practice

20.37   Having escaped to the sea, survivors have to be rescued and taken to a safe place where medical care is available. The means of rescue should be effective in the widest possible range of weather conditions. Reg 10 of the Emergency Procedures Regulations requires a SBV to be present within 5 nautical miles of every normally manned installation, ready to give assistance in the event of an emergency at or near the installation. These vessels carry FRCs which can be deployed rapidly to rescue persons from the sea. This regulation is supported by a DoT code of practice, for the assessment of the suitability of SBVs for attending offshore installations, setting out standards which are to be met as the condition of the issue of a certificate in respect of the vessel. The code was first prepared in 1974; its binding force is based on a voluntary agreement whereby members of UKOOA undertook to abide by the standards set out in the code. Mr B C Drew, a senior surveyor in the DoT, said that this voluntary agreement had been honoured; to his knowledge there had never been a case where a vessel which has not been certificated has been chartered by a member of UKOOA. The third and latest edition of this code was issued in draft in August 1989. It took account of lessons learnt from the disaster. It has not yet been ratified.

Requirements of the code of practice

20.38   Mr Drew explained that the first code of practice continued in use after the introduction of the Emergency Procedures Regulations.

The latest ratified version was issued in 1984. It required, *inter alia*:

1. Provision of FRCs.

2. Improvement to the accommodation.

3. Improvement to equipment and first-aid.

4. The provision of bridge control.

5. Special provisions for large SBVs and those operating north of 62° north and west of 15° west.

Work on the third edition was started in 1986. In it the functions of an SBV have been expanded to include the requirements that it should:

(a) Communicate with the installation, etc.

(b) Rescue persons from the water.

(c) Keep a monitoring watch in the safety zone.

(d) Attend closely to the installation during certain operations (helicopter movements, work over the side, etc).

As stated in the preceding paragraph, this edition has not yet been ratified.

Criticism of SBV legislation

20.39 UKOOA witnesses were critical of the existing legislation governing the requirements for SBVs. Mr C J Middleton, Chairman of the UKOOA Marine Committee, said that the SBV was only one part of the total evacuation and rescue package but nevertheless a whole regulation and a code of practice were devoted to it. The prescriptive requirements that the SBV should have accommodation on board for the total population of the installation and that the SBV should attend each individual installation; and the specification of vessels by broad geographical areas were particularly criticised. It was suggested that the regulations should instead require operators to propose a total evacuation and rescue package for each installation or group of installations. Mr D T Rudd, of BP, described BP's evacuation policy and plan for the Forties field. This was an assessment of the total evacuation and rescue requirements for the Forties complex of 5 fixed installations and proposed that 3 SBVs (or 2 SBVs and the emergency support vessel Iolair) would be adequate to ensure safe and full evacuation and rescue in an emergency in that complex; or 4 SBVs when a flotel was stationed in the area. This contrasted with a requirement of 8 SBVs if Reg 10 was interpreted in a strict manner, as the DEn had recently indicated should be done. BP presently employed 4 SBVs in the Forties field. Mr Middleton suggested that a remote drilling rig, say to the west of Scotland, which would have little back-up nearby, would require a high-capability SBV which could accommodate the total population but such vessels would not be required in a multi-platform field or in a developed area of the North Sea. The consideration of a total package of facilities and other arrangements for evacuation, escape and rescue rather than a simplistic prescription of standard requirements is in line with the approach which I have already recommended in this chapter (paras 20.8-9). I recommend that the required changes are made in the legislation and code of practice so that evacuation, escape and rescue can be the subject of an analysis submitted by each operator; and form part of the Safety Case for each installation. However, some prescriptive regulations on standards and quality of equipment, crewing and training would be required. These equipments and standards are discussed below.

Criticism of SBV standards and the UKCS SBV fleet

20.40 Criticism by survivors and other witnesses was levelled at the general standard of SBVs in the UKCS. Information provided by Mr J S Daniel, Chairman of the Standby Ship Operators Association, showed that of the 187 SBVs for which there were complete data, 162 (87%) were converted fishing trawlers. There were only 7 purpose-built SBVs and 18 multi-functional vessels operating in a standby mode, but not all full-time. Mr Daniel said that if fishing trawlers were properly converted they would meet the requirements of the code; a trawler hull provided a very good platform for rescue operations because of its good sea-keeping qualities. However Mr M Macey, a Director of Maritime Rescue Services Ltd, was doubtful whether trawlers were sufficiently manoeuvrable even with bow-thrusters fitted, or had the visibility to pick up persons easily from the sea because of their small bridges and high forecastles. Mr Middleton suggested that because of the requirement in the code that SBVs should be capable of accommodating the total population of an installation, innovation had been stifled and the use of aged trawlers had been perpetuated.

20.41 The only SBV about which the Inquiry heard detailed evidence was the *Silver Pit*. I described its performance on the night of 6 July in Chapter 9 in critical terms.

It was deficient in many respects, although the courage of its crew was outstanding. In the light of the evidence my impression was that a large part of the 162 converted trawlers in the UKCS SBV fleet were in no better position. Offshore operators were accused of taking the view that because SBVs make no contribution to profit, expenditure on them should be kept to a minimum, a view vigorously resisted by UKOOA. Mr Daniel said that the pressures put on charter rates after the fall in oil prices in 1985/86 had made it difficult to maintain standards. The Norwegian standby fleet is in large part purpose-built to the specifications required by Norwegian legislation. My understanding is that the 7 purpose-built SBVs operating in the UKCS waters were mainly, if not all, Norwegian-built. I accept the implication that the strictness of the regulation and the present code have discouraged some operators from doing other than the minimum necessary and have thus inhibited improvement of the SBV fleet; and that the cost of operating SBVs has not necessarily enjoyed a high priority in the operating budgets of oil companies in recent years. I strongly urge that the standard of the existing SBV fleet is improved with despatch, although it is obvious that this cannot be done at once. Basic standards should be introduced for existing vessels and a tight but realistic deadline for compliance set. Specifications for new vessels should be set which will ensure that they meet fully the requirements of the Safety Case.

SBV equipment quality and standards

20.42   In these circumstances there is a strong case for setting specific standards for SBVs in legislation to ensure that the vessels used are of consistent quality and reliability. The important mechanical standards for SBVs, apart from sea-worthiness, should include:

1. *Manoeuvrability* It should not be a requirement for an SBV to manoeuvre close to damaged installations. Rather it should maintain a safe distance and use its FRCs to pick up survivors from the installation if this is possible. It should be able to manoeuvre to pick up survivors from the water or clinging to wreckage. This requires it to be highly manoeuvrable and able to maintain its position.

2. *Visibility* The master should be able to keep the rescue areas in full view from the bridge, and be able to approach a person or object in the water while retaining total control of the vessel. The FRC launching area should be fully visible from the bridge.

3. *Lighting* At least 2 searchlights, covering the full 360°, and capable of being remotely controlled from the bridge, should be available. There should also be adequate local lighting in the pick-up and launching areas.

4. *Communication* The SBV must be able to contact its FRCs, the installation and nearby vessels and aircraft, as well as to maintain conversational contact between the master and the crew.

5. *Survivor recovery from the water* Scrambling nets or ladders are only suitable for use by fit and uninjured personnel. If an accident has caused the escape of people to the sea it is very likely that some will be injured, possibly severely. A number of recovery techniques were described to the Inquiry. At least 2 different methods other than ladders or scrambling nets should be available on the SBV.

6. *FRCs* Two FRCs should be carried on SBVs. FRC standards are discussed in para 20.46.

7. *FRC launching and recovery* A rapid launching facility for FRCs must be installed. Launching should not be in the critical path of emergency response, ie the FRC should be able to be put in the water with engine running as soon as the crew is ready. Consideration should be given to the need to recover an FRC with a badly injured survivor on board who would be in danger by conventional recovery.

These should apply independently of what may be shown by the Safety Case to be required: and accordingly should be prescribed by regulations which would otherwise be goal-oriented with regard to rescue facilities.

SBV usage in man overboard incidents

20.43   In practice the main use of SBVs has been in the rescue of men falling overboard from an installation, either when working over the side (scaffolders etc) or other accidents (knocked overboard, etc). Where men are working over the side, SBVs or larger FRCs are stationed in close proximity to the installation. Data presented to the Inquiry showed that 126 MOB incidents occurred in the 13 years from 1974 to 1986 inclusive. The most common means of recovering the MOBs was by FRCs deployed from SBVs. A total of 35 deaths resulted from these MOB incidents. An analysis of this data showed a survival rate of 95% for working over the side incidents, compared with survival rates of only 45% on other installation MOB incidents and 60% for MOBs from attendant vessels (the latter including 7 from SBVs themselves). Over the period there were an average of 9 MOB incidents per year involving 10 or 11 casualties of which approximately 3 were fatalities. In the 26 years since North Sea operations started in 1964, there has been only one incident in which men had to escape into the sea deliberately - namely the disaster. This emphasises the need to ensure that the requirements of the regulations allow for MOB incidents and do not concentrate solely on major disasters which are rare.

SBV usage on other duties

20.44   The code for assessment of SBVs indicates that an SBV can operate in a multi-purpose mode provided that its safety role takes precedence over all others. Mr Drew explained that, for example, a supply vessel with appropriate standby certification can carry cargo from port to an installation provided that sufficient cargo is unloaded and the rescue areas are maintained clear of all encumbrances before she becomes a dedicated SBV. Also, certain small quantities of cargo can be carried between installations without impairing the safety function of the vessel. It is for the master of the SBV to ensure that the vessel is ready to undertake the rescue role at any time when on standby duty. Mr Ognedal also explained that in Norway SBVs could undertake functions other than standby services provided that these did not hinder the standby tasks or affect response times. This does not appear to be an area of difficulty.

FRC equipment and performance standards

20.45   The code for assessment of SBVs lays down that at least one FRC is carried on SBVs, ready for immediate use and capable of carrying at least 9 people plus 2 crew, or 15 people and 3 crew in the larger specification. FRCs must be capable of being launched while the SBV is underway. Their engines can be either petrol or diesel driven and be capable of being maintained while the SBV is on station. Although not specifically recommended in the code, all FRCs are of the rigid hull inflatable type and are self-righting. The larger ones are fitted with an enclosed wheelhouse which allows their crew to continue on station for extended periods, as required by MOB and other duties of attendance. They are capable of a speed of at least 20 knots in average sea states.

20.46   The lessons drawn from the disaster show that the FRC is a very important part of the total rescue effort. It is particularly important that they are fully reliable mechanically. I consider that it is very important that there should be a very high standard in a number of areas including:-

1. Launching capability
2. Capacity
3. Speed
4. Crewing
5. Maintainability
6. Communication links
7. Search capability (lighting etc).

FRC reaction time

20.47   Evidence was heard that when someone fell into the North Sea in winter there was the danger that he might suffer from "cold shock", a stoppage of the heart. If he was recovered within 4-5 minutes he stood a chance of recovery without permanent brain damage. If he survived the initial immersion the average period he could survive before hypothermia caused death was about 30 minutes, but this varied considerably from person to person. No target times for MOB recoveries have been recommended in the code but it is accepted that in over-the-side work supervised by SBVs and/or FRCs a recovery time of less than 5 minutes is achievable. Captain Ginn said that it would be very difficult to match that performance by helicopters, which would have to be on the same platform, fully manned and equipped with search and rescue facilities. Data from the period 1974-1986 showed that of the 43 casualties recorded as being in the water for 1-5 minutes, only one died (2%). Of 14 casualties in the water for 6-10 minutes, 2 died (14%) and of 14 in the water for greater than 10 minutes, 5 died (36%).

FRCs on installations

20.48   FRCs can be deployed from offshore installations as well as from SBVs. This is common practice in the NCS where 48 rigs and platforms carry FRCs. They are installed on only 3 installations in the UKCS. Mr Wallace said that the industry was not encouraged by the regulations to install FRCs on installations; FRCs were required on the SBVs in attendance. However, if the regulations encouraged a flexible approach rather than rigid prescription, in the case of a cluster of installations FRCs on some installations could give better cover. They would be particularly useful in covering work over the side. There were problems with launching and recovery. These problems have apparently been overcome in the NCS although this took about 8 years. I would recommend that the opportunity to station FRCs on installations should not be constrained by regulation, as their use would probably be an attractive part of total evacuation and rescue packages for installations (see para 20.39).

20.49   FRCs are launched from installations either by using single-fall davits (normal in Norway) or by cranes, with or without special launching cradles. It was recommended by Mr Wallace that they should be mounted as low as possible, ie on the cellar deck (the module support frame) as they did not have to be sited at emergency embarkation points. Dr Side reported that the reaction times for MOB recoveries using installation-launched FRCs compared favourably with those from SBVs during periods of work over the side. The longest response time found in a study of MOB incidents was 7 minutes, with the average being 4-5 minutes. It might not be necessary to have a dedicated FRC crew available on the installation as they might be able to carry out other non-conflicting duties when not required for MOB cover.

SBV and FRC manning levels and requirements

20.50   The code for the assessment of the suitability of SBVs for attending offshore installations requires that the DoT states on the certificate of survey of the SBV the absolute minimum manning below which it is considered unsafe to operate. The owner of each vessel in association with the master prepares contingency plans covering the responsibilities and allocation of duties of crew members in the event of the occurrence of an incident. These plans also detail descriptions of the responsibilities of the SBV master and the installation OIM, and the responsibilities for the control of search and rescue operations. The minimum manning scales range from a crew of 7 to a crew of 15 depending on the size (ie survivor capacity) of the SBV. Mr Macey felt that the manning specifications in the code were too low with respect to the smaller size SBVs. Also when 2 FRCs were carried there should be crews for each of these on board (3 each). This was not allowed for in the code. It was also insufficient in not requiring a second mate, to alleviate over-long watches. I do not consider that the evidence before the Inquiry was such as to enable me to make definite recommendations on this matter.

I consider that the DoT should take this evidence into account when revising the code.

20.51 The proposals in the amended code require medical examinations before employment for all members of the crew. Crew members over 50 have to undergo annual medical examination, and be certified as fit to be employed on a vessel offshore for up to 28 days at a time. Except in special cases the age limit for crew members should be set at 60. I agree with these proposed standards.

20.52 The proposals also state that the periods of duty on SBVs should be limited to 28 days in summer and 21 days in winter, with an allowance of 2 days steaming to and from installations. No crew should serve 2 consecutive periods of duty with less than one week's leave of absence between them. Again I am in agreement. I consider that the changes referred to in this and the last paragraph should lead to an improvement in performance and the enhancement of safety.

SBV crewing problems

20.53 Low charter rates in an increasingly competitive market in recent years appear to have made it difficult to find and retain people with the appropriate knowledge, experience and mentality to crew SBVs. Mr Macey suggested that operating companies did not take sufficient interest in ensuring that crew standards and training were up to requirements; and that crews overstayed their prescribed tours of duty. (Mr Drew of the DoT said that this latter problem was being acted on.) Mr Daniel suggested that better conditions, more training and better pay (which would entail higher charter rates) would be needed to improve the manning situation. He said that Government, operating companies and vessel owners/operators should seek to engage in constructive discussion so that improvement can be achieved and fragmentation avoided.

20.54 The problems of crew motivation and boredom, and their feeling of being unappreciated by the installation and the industry were described by Mr Macey. He suggested that more contacts between SBV crews and operating company personnel, both offshore and onshore, would help. So also would more regular offshore exercises involving the SBVs and their crews. The different terms and conditions of SBV crews, who did not enjoy the equal time on and off duty that was common in operating companies, did not help their motivation. Crew accommodation and recreation facilities were obvious areas in which to seek improvements but stimulation by activities, exercises and close involvement with the platform were also important. It would appear to me that there are indeed real problems with the motivation of SBV crews who have the task of keeping station for weeks at a stretch with nothing to relieve the routine. This is not a matter for regulation but I consider the offshore operators and vessel owners should take steps to improve the situation. As a minimum there should be more contact between the SBVs and the installations and more realistic exercises. In this respect I would recommend that the position and status of SBVs offshore and their functions for the following week should be notified weekly to the regulatory body with a copy to the DoT. This would minimise the possible over-staying of tours of duty and would also keep the situation of the crews in focus for vessel owners and charterers.

SBV crew training

20.55 There is no current legislation covering the training of SBV crew members. The latest edition of the code sets out the certificates which have to be obtained by all crew members and gives guidance on the establishments and bodies where the training courses specified can be undertaken. Requirements for specialised training are laid down. Every member of the crew should have attended a course of basic first-aid. At least 2 of the crew (other than the master) should hold a certificate in advanced first-aid, one of whom should be nominated as the medic. For each FRC carried at least 3 crew members should be trained and hold certificates in all aspects of its handling and communications. Refresher courses must be attended regularly. The courses specified in the code appear to cover all the recommendations for training made in the course of evidence. The government departments, vessel owners and operators and operating companies should co-operate to ensure that all crew are fully trained in all aspects specified. I would emphasise the need to train and refresh the crews of SBVs, especially the coxswains and crews of the FRCs. All training should be documented and records of training held centrally, preferably by the OPITB. Probably the most important concern must again be to motivate the crews to take the benefit of the training. As Mr Macey, an expert on training, put it, "It does not matter how much training you give the crews, if the crew is not motivated then you are wasting your time.

Command in emergencies

The command structure

20.56   Evidence was heard from UKOOA on the command structure and organisation required to ensure effective response to an offshore emergency. Mr M R Baxendine, a Shell Expro OIM, said that in his company the general practice was to pre-select, train and drill at least 3 emergency response teams (operations, drilling and services), all with trained back-up support teams available in case these were required. All non-essential personnel, (ie those not in the command structure or the emergency response teams), assembled at predetermined muster stations, grouped by the lifeboat numbers to which they were assigned on arrival on the installation. The OIM was in overall charge of the installation; his replacement was pre-nominated to replace him if he was incapacitated or not contactable. Only the OIM or his replacement had the power to decide whether it was necessary to abandon the platform. His emergency command centre, normally the radio room, would receive progress information from the response teams and would direct them and communicate with nearby vessels and the shore as necessary. This appears to be the general pattern of emergency command followed on the UKCS. I comment on the criteria for OIM selection in the next paragraphs.

The OIM

20.57   The appointment of an OIM is required under the MWA (Secs 4 and 5). The regulations give the OIM general responsibility for safety, health and welfare and for maintaining discipline and order on the installation. Candidates for the post are nominated to and accepted by the regulatory authority. The Inquiry did not hear any evidence on the criteria applied to the acceptance of an OIM by the regulatory authority. In earlier chapters I expressed the view that there were significant shortcomings in the performance of certain of the OIMs on the night of the disaster. In particular I expressed the view that because of the lack of leadership on Piper the death toll was substantially greater than it would otherwise have been. A number of survivors said that in the galley, where the OIM was positioned, no one was in charge or giving instructions or advice.

20.58   Evidence on the abilities required in an OIM was given to the Inquiry by UKOOA and by the Institution of Mechanical Engineers. Mr K A J Ellice, a training manager with BP Exploration, said that they looked for exposure to the North Sea environment, experience in a related technical discipline and ability to command. Information on ability to command would normally be provided through the in-company staff appraisal systems. It was extremely difficult to judge a person's ability to command in a precise way; they could be provided with the "techniques and mechanisms". He said that leaders were found rather than trained. Mr Ellice was definite that BP would not and do not use psychological tests such as were practised in the Royal Navy and Merchant Navy. During experience and training leading up to the selection there would be the opportunity to assess individuals in a variety of situations and circumstances. Dr A A Denton, representing the Institution of Mechanical Engineers, listed 4 criteria that the OIM and at least one deputy should have, command ability, specifically tested in simulated circumstances; technical literacy to at least Higher National Diploma or equivalent standard; experience of at least 3 years offshore; and understanding of the sea/air environment by training and experience. Mr Baxendine, who gave separate evidence on the command structure, was a practising OIM with 14 years experience. He stated that in the vast majority of cases the OIMs in his company (Shell) had all previously commanded groups of men.

20.59   The failure of the OIMs to cope with the problems they faced on the night of the disaster clearly demonstrates that conventional selection and training of OIMs is no guarantee of ability to cope if the man himself is not able in the end to take critical decisions and lead those under his command in a time of extreme stress. While psychological tests may not appeal to some companies the processes used and proven successful by the armed forces or the Merchant Navy, who have to rely on their officers to lead under stress, should be seriously considered by operating companies. The post of OIM calls for decisions which may make the difference between the life and death of personnel on board. The remoteness of installations, the requirement for installations to be self-contained in the means of dealing with a rapidly developing incident, the need to obtain, verify and consider data communicated to him from various sources for immediate decision on which the lives of those on board depend demands a level of command ability which is not a feature of normal management posts. The command ability of the OIM and the command structure and organisation in emergencies should be factors in the Safety Case proposed by the operating company. They should be part of the safety management system of the company which I will propose in Chapter 21.

Emergency exercises

20.60   Mr Ellice described how operating companies tested their command structures, by regular emergency exercises held for each installation, by operating company exercises and by full-scale exercises involving outside authorities such as the coastguard, police etc. For these a major disaster such as a helicopter crash on a platform or an explosion and fire were simulated. BP employed emergency response trainers who regularly visited all installations and assisted installation management to conduct specialised emergency exercises. There were also nominated persons in their safety department who were totally responsible for the planning and instigation of large-scale emergency exercises carefully organised to require co-ordinated onshore and offshore response. They considered that full-scale exercises were very important, covering interconnected platforms as necessary. Larger-scale exercises effectively exercised the emergency systems as well as training those in command on the installation. UKOOA have published guidelines for offshore emergency exercises, to determine the effectiveness of the operators' emergency procedures. Both in-house and major exercises in conjunction with outside authorities are specified.

20.61   I consider that emergency exercises are essential means of ensuring that paper procedures work in practice. They also allow for the assessment and upgrading, as necessary, of the performance of the command structure. I recommend that emergency exercises are carried out in accordance with UKOOA guidelines and that command teams are given practice in decision-making. The operator's system for emergency exercises will form part of its safety management system (see Chapter 21).

Precautionary musters, drills and training

20.62   Precautionary musters are held on all North Sea installations. All the persons on board are assigned a TEMPSC number on arrival on the platform and, on the emergency alarm sounding, non-essential personnel assemble at the pre-determined muster station, by lifeboat groups. If the OIM decides that the platform should be wholly or partly evacuated they make their way either to the helicopter deck, if helicopters are used, or to the lifeboat station, where pre-selected coxwains command the individual lifeboats. Separate lifeboats are reserved for the emergency response teams to use if it becomes necessary to evacuate the installation completely. POB lists are maintained on installations but experience on the night of the disaster shows that it would be important for them to be updated for every movement of personnel, and copied immediately to shore. These lists should be maintained in alphabetical order and by contractor employer, to minimise confusion and delay in reacting to queries in emergencies. I recommend that all POB should attend at least one muster per tour of duty; and that the circumstances of all precautionary musters and evacuations should be reported to the regulatory authority.

20.63 If central control and a planned evacuation cannot be exercised, as in the case of Piper on the night of the disaster, the personnel would be expected to make their own way to the sea. Mr D S Kinloch of Conoco described the need to take individual action in these circumstances. Such action, if taken prematurely, could of course be detrimental to controlled and orderly evacuation but the emergency training given to offshore personnel should enable individuals to minimise the risks they take if this becomes inevitable.

20.64 Realistic and up-dated emergency training and regular drills are of vital importance to ensure that the risks of emergency evacuation, escape and rescue are minimised. They should never be neglected. I recommend that the UKOOA guidelines for offshore emergency safety training on installations should be a minimum requirement for emergency and related training. I recommend that records of personal details and safety training courses attended by all personnel seeking employment offshore should be maintained by operators until the central training register instituted by the OPITB is operational. As for emergency drills and training, the operator's systems for these should form part of the safety management system (see Chapter 21).

20.65 The responsibility to ensure that the reaction to an emergency is effective, safe and disciplined is primarily with the management of the installation, ultimately the OIM. Onshore management have important roles to play in this but the decisions have to be taken on the platform. The command structure must be tested and drilled regularly. This should be seen as an essential part of working offshore. The biggest difficulty arises where there is an attitude of indifference offshore, particularly in the case of the occasional worker who may be on the platform for only a few days. All on board should take part in training, drills and exercises. The abnormally high casualty rate among those on Piper who, for reasons of their employment were not fully familiar with the platform layout, was striking.

Chapter 21

The Future Offshore Safety Regime

Introduction

21.1 The discussion in this chapter is divided into 3 main parts in which I discuss:

(i) The importance of the management of safety by operators; and the need for the maintenance of a consistently high standard of performance if their responsibilities under the regime are to be discharged (paras 21.2-14),

(ii) the extent to which the present methods of control, allocation of responsibilities, regulations and guidance in the offshore regime are appropriate and effective (para 21.15-51), and

(iii) changes in the regime which, along with those recommended in Chapters 17-20, are in my view necessary if the regime is to fulfil its functions in an appropriate and effective way (paras 21.52-87).

The management of safety

The role of operators

21.2 The safety of personnel on installations is critically dependent on the management of safety by the operators, as the circumstances of the disaster clearly demonstrated. There is, of course, nothing new in the idea that safety requires to be managed. The reports of investigations into recent major incidents have shown the dangers posed by serious failings in the management of safety by large organisations.

21.3 The evidence before the Inquiry has served to demonstrate that an offshore installation presents a combination of features which make it unique from a safety point of view. The living quarters are relatively close to the plant, which itself is placed in a confined space. Evacuation may be difficult if not impossible in certain weather conditions. While a chemical plant has some of these characteristics, evacuation is always available and the operating crew are not confined to the immediate vicinity. Installations are designed to meet specific requirements and may be subject to modification. These considerations underline the need for an adequate system for the management of safety and the need for a suitably rigorous regime which ensures that this is maintained.

21.4 There are practical limits to the extent to which a safety regime can affect the manner in which safety is managed. Mr R E McKee, Chairman and Managing Director of Conoco (UK) Ltd, offered the following comments:

"It is my fundamental belief that safety cannot be legislated, while recognising that enough legislation or regulation needs to exist to ensure that minimum standards are maintained. Such regulation should impose a duty on the operator to do everything reasonable to achieve a safe operation. By and large, safety has to be organised by those who are directly affected by the implications of failure. These people are in the best position to determine the detailed measures necessary on their own particular installation to achieve the safety objective. Imposition of detailed requirements cannot anticipate all the variances of differing practice, location, organisation and size that exist. In fact, prescriptive regulation or over-detailed guidance may at times result in the overall safety objective actually being compromised. Innovation, on-going improvement and objectivity will be stifled; and the more prescriptive the regulation the more unclear it is who has the responsibility for total safety. Compliance becomes the overriding objective. Sight is lost of the more realistic and overall intent that all reasonable steps should be taken to achieve the total safety of the installation. Finding the middle ground is difficult. The

Government is faced in some ways with the same problems that upper management is. In other words, first, they must be confident that industry has in place facilities that are properly designed for safety, using a FSA approach, and then that organisations have generated a proper safety culture that will help result in excellent safety performance. Audits need to assume a far higher prominence as a means of checking the ability of the organisation to achieve safe designs, operations practices and systems to interrupt a chain of events leading to a Piper Alpha type accident. This will require more skilled personnel for operations to conduct specialist audits, for third parties to check them and for Government departments to review their success."

With the general thrust of those comments I find myself entirely in agreement. I should add that it is also plain that a regulator cannot be expected to assume direct responsibility for the on-going management of safety. There may be circumstances in which inspectors can and should take the relatively drastic step of interfering by means of statutory notices, but these are the exception. For all practical purposes the management of safety is and remains in the hands of the operators.

21.5   This approach may be compared with what Mr Rimington described as the HSC's approach to the principle of self-regulation:

"For practical purposes, its essence is that while the regulator can and should, in consultation with those regulated, provide a framework of rules and the necessary impulses and disciplines, health and safety is principally a matter for management in-firm."

21.6   These considerations underline the importance in the offshore safety regime of the general duties of employers under the HSWA and measures which are directed to ensuring that these duties are performed in a demonstrably adequate manner. I regard this as a key part of the regime.

Changes in the regime discussed in earlier chapters

21.7   In Chapters 18-20 I have discussed a series of measures which should be put into effect in the interests of safety, primarily through goal-setting regulations and to some extent through more detailed provisions.

21.8   More fundamental than these, and different in kind, are the measures directed to the submission and acceptance of a Safety Case on certain aspects of safety which I have discussed in Chapters 17, 19 and 20. These are directed to ensuring that the potential major hazards of the installation and the risks to personnel thereon have been identified and appropriate controls provided; and that adequate provision is made for ensuring, in the event of a major emergency affecting the installation, a temporary safe refuge (TSR) for the personnel on the installation, and their safe and full evacuation, escape and rescue. The present chapter will consider whether further requirements should be imposed on operators with a view to their demonstrating that they have made and maintain adequate arrangements for the management of safety at large.

The means of achieving adequate management of safety by all operators

21.9   Management witnesses gave evidence to the Inquiry as to the importance of defining and communicating to the whole workforce an adequate safety culture or philosophy; and ensuring that they were fully motivated to implement it. Safety should not be treated as something which was separate from the conduct of business. Mr R A Sheppard, Vice-President of Production and a Director of Amoco (UK) Exploration Company said that "safe, prudent working practices and procedures are good business practices". The organisation of safety was a matter for line management at each successive level. Mr McKee said:

"Philosophically we look to line management for safety performance, not to the safety department or to a government agency. If a safety programme is to have outstanding results, it is imperative that senior and then each progressively junior level of management exerts its leadership in establishing goals, demanding accountability for performance and providing the necessary resources. While top management sets the safety standards for the entire company, our first-line supervisors are the key link in actually making it happen. Each of them is personally involved in safety training, safety inspections and other safety activities. They make sure that all line employees and contractors reporting to them are trained to work safely: that not only must they know how to perform their jobs properly and safely but are convinced that they have a responsibility to do so."

21.10 Mr Rimington pointed out that the reports on recent major incidents had drawn attention to the importance of the chain of command for safety, and particularly to the significance of leadership from the top. They had also focused attention on the related aspect of in-firm safety culture, and particulary the influence of the human factor in accident causation. That close attention to the management of safety was effective in preventing accidents, and that it was compatible or even associated, with first rate commercial performance was clearly demonstrated. For this he cited the performance of Du Pont de Nemours (of which Conoco (UK) Ltd is a subsidiary). His inspectors had formed the unequivocal impression that the more successful firms usually adopted a more highly structured and effective approach to safety than others. He held the view strongly that it had a great deal to do with discipline. "If one adopts a disciplined and determined approach to one's commercial success, one is likely to adopt such an approach to other aspects of one's business." The establishment of a safety culture included, he said, the "systematic identification and assessment of hazards and the devising and exercise of preventive systems which are subject to audit and review. In such approaches particular attention is given to the investigation of error. The control of human error involves the assumption that people will make mistakes but that by thought, pre-design and proper motivation this can be made much more difficult and the consequences mitigated."

21.11 It is clear that a systematic approach is required if an operator is to ensure safety on its installations and compliance with the requirements of legislation, including the duties imposed under the HSWA. This involves a planned approach to the elimination of danger both in design and in operation. This may be illustrated by the evidence of Mr P Doble, Deputy Project Manager of the Kittiwake Project, who explained how its design proceeded through the stages of feasibility study, conceptual design and detail design. The design philosophy was documented so that in any subsequent audit or review it would be possible to judge what had actually been done against the design intent. Management procedures were based on the quality assurance specifications of BS 5750, supplemented by systematic hazard identification and analysis to provide a series of checks and balances as the design proceeded. These included hazard and operability studies, safety reviews, equipment criticality assessments and audits. Mr McKee said that Conoco traditionally had a formal quality assurance system in place for design and construction of new platforms and modifications to existing facilities. The extension of this philosophy to the technical aspects of their operations was a key objective for 1990. The objective was to have a regularised way of watching over work practices so as to interrupt the chain of events that resulted in an accident. Mr M Ferrow, Manager, Safety and Quality Assurance for Conoco said that safety could be regarded as a "sub-set of quality assurance". He described how quality assurance was used to "close out" study findings in the engineering safety plan for Conoco's "V Fields" in the southern North Sea. He believed that it would be impossible to have a fully safe operation without a quality assurance system which he described as "a mechanism by which managers and engineers and the company in general can be sure that what they are doing will be safe, operable and fit for purpose, and, bearing in mind the faults and errors which can occur, either technical or human, sets up systems which take reasonable precautions against that; and then, finally, imposes some sort of formal audit structure into that to ensure that these things are being done on a continual basis." The use of quality assurance had evolved in the North Sea industry from the early or middle 1970s, starting with structural matters, extending to systems and then into an all-embracing technique for ensuring that what was designed was in fact built to specification. It was now moving, or had moved, into the operating areas.

21.12   Mr Rimington described quality assurance in the promotion of safety as "absolutely essential" and established practice in all major industries. Mr Petrie was aware that most operators demonstrated some type of quality management system to a standard such as BS 5750 or ISO 9000.

21.13   Common sense and experience of what happened on Piper indicate that it is not enough to set up a systematic approach to safety and put it into operation. There is a plain need to review and up-date the system in the light of experience both of the operator and of the industry. It is also necessary to "audit" the extent and quality of adherence to the system and to "verify" that its results are in practice satisfactory. It is clear that companies with an outstanding safety record go to considerable lengths to audit the management of safety. Mr McKee explained that Conoco performed management and safety department audits and inspections on a frequent basis, amounting to several per month on all aspects of safety management from housekeeping to work permit usage. As chief executive he received safety audit reports and reacted by raising the issues which were involved with the relevant vice-president. He conducted his own informal audits on his frequent visits to the platforms. He specifically discussed the audit system with managers and employees. All operating managers and their staff conducted regular internal safety audits. First line supervisors conducted frequent safety inspections in their areas of responsibility. Company procedures required daily audits for compliance with the PTW system and weekly compliance reports. Experienced safety professionals carried out audits on a regular basis by means of inspections of the platforms, including the activities of contractors. In addition there was an annual management safety audit of the platforms, including all relevant onshore managements. There were also special safety audits including team members from outside consultants. Mr Sheppard observed that "simply looking at the way the equipment is operated and the operating conditions of the equipment is not a complete safety audit. It has to incorporate in it the operating procedures, the way safety is approached, upon that particular installation, and it cascades up into how it is approached generally by the company." If it was found that part of the safety philosophy or safety programme was not being followed or interpreted in the appropriate manner he would discuss his concern with the manager of the particular part of the company to find out whether he had the same impression. Contributions might also be sought from the safety specialists and other Amoco managers in order to see whether his assessment was confirmed. Together they would examine the available data and determine whether they were dealing with the root cause of the problem and not a symptom of some larger problem. Once that examination was complete they would collectively set about exploring ways of improving the situation. It could be a communication problem, a supervisory deficiency, a training issue or an organisational flaw.

Relationship to the regime

21.14   The need for a good and well maintained management of safety by all operators, and not merely the few, is plain. No doubt the prime responsibility for this lies with the operators. However it is also plain that the regime has a part to play in the achievement of that overall objective. Is the present regime appropriate and adequate for the purpose? I turn next to this question.

The existing regulatory approach

21.15   In this part of the chapter I will examine the present methods of control; the allocation of responsibilities; and existing regulations and guidance.

Methods of control - design

21.16 Since the Construction and Survey Regulations came into operation the examination of design and what has been constructed has effectively been in the hands of the certifying authorities. The Second Schedule to the regulations coupled to the guidance provided by the DEn require certifying authorities to consider whether various aspects of an installation meet and continue to meet specified standards, frequently related to establish codes. As I explained in para 16.27, certifying authorities are concerned with conceptual design of process plant only to a limited extent. In particular they are not required to review plant design in relation to major hazards. Their concern is with the end product or the proposed end product of the design or construction. They are not required to examine the management systems which lead to that design. As I have already stated in Chapter 19 while certifying authorities are concerned with passive fire protection, active fire protection is the concern of another body.

21.17 Under the present regime there are no other requirements which oblige operators to show that their management of safety is adequate for the purpose of safety in design.

Methods of control - operation

21.18 It is clear from the evidence that the DEn take the view that it is essential that the quality of management is assessed by them and found adequate. Mr Petrie's position was that over the years, as part of inspection activities and other activities with companies this assessment of management had occurred. He enlarged on these other activities by saying that there were management safety presentations "which start with the senior management of the company, where they describe their philosophy for safety and their control of safety in management terms, and how they implement that right through their structure to the relevant people on the installation - the managers, supervisors and other staff." That was one element. "From that follow discussions at different levels with my people in their appropriate levels within the company, and it is finally down to the assessment of inspectors in undertaking offshore inspections." He pointed out that the monitoring of management of safety was very similar to what was done onshore. His department had carried that forward with the concept of FSAs, which brought in all aspects of safety.

21.19 The PED's programme for 1988/89 which was submitted to the HSE in November 1987 shows that until the first half of 1987 the PED's ability to undertake safety presentations had been severely constrained by widespread over-loading of the staff in the safety branches, particularly at management level. What actually was done by way of safety presentations did not emerge clearly in the evidence, despite the attempts of various counsel to obtain elucidation. What seems to have happened is that presentations by senior management did not take place but that a number of companies gave presentations at middle management and working levels. As regards the slippage Mr Petrie explained that safety presentations involved a significant workload for his department and for himself in particular. However, prior to the disaster his department had carried out a pilot study based on presentations by a cross-section of companies, about 8 in number. This was, as Mr Petrie put it, "to see the way ahead and to see if it was of any benefit". The study had shown that the presentations, he said, "could be an effective method of assessing as well as stimulating operators' management commitment to safety, as well as providing valuable information for inspectors engaged in offshore inspections. A section to co-ordinate and service such audits in a way consistent with the safety management system requirements of the regime is being set up. Advice has been sought from the HSE's accident prevention and advisory unit and from others such as NPD on the audit of safety management systems." The disaster had caused the cancellation of further presentations. They would probably be reinstated in the Spring of 1990. Additional resources had been agreed for that.

21.20   I am bound to say that I see the system of inspections of the conventional type practised by the PED, when considered as a means of assessing or monitoring the management of safety by operators as suffering from a number of fundamental limitations. While inspections may lead to correction both at the time and for the future they address something which has already gone wrong. There is no systematic examination of the operator's system for the management of safety. In particular an examination of management onshore is not involved unless something comes to the attention of an inspector during a comparatively short offshore visit (reference may be made to para 15.50.). The obtaining and following up of safety presentations would be a means of the PED coming to grips with operators' safety management systems, but the progress in that direction has been extraordinarily slow and tentative. In any event it is still no part of the requirements of the regime that a safety management system should be demonstrated to be adequate and carried out in practice.

The PED's approach to the future

21.21   In considering the existing regulatory approach I should also take account of the regulatory body's approach to the future. Mr Petrie said that amongst the most important improvements that could be carried out in the regime in the next couple of years included taking forward the FSA approach as fast as possible and having it considered by operators. Another was the move towards goal-setting regulations, replacing prescriptive regulations which were over-inhibiting. Mr Priddle said that the most important area which he would identify for the future was that attention and a new focus required to be given to the development of the capability of the Safety Directorate to assess management systems.

21.22   Mr Petrie agreed that while no regulatory body can expect to employ permanently the full range of expertise which may be required for its work it ought to have most or all of the expertise needed to cover continuous requirements of its system of regulations. Within the PED there were people who had reasonable knowledge of management systems but assistance was being sought from the HSE. Mr E J Gorse who spoke to the DEn's discussion paper on FSA, made it clear that in the absence of in-house expertise the PED were employing consultants in addition to assistance from the HSE in formulating the Department's proposals. Mr Priddle said that while the Safety Directorate had a general competence in relevant engineering disciplines, he knew that there were some particular specialist disciplines which were not represented. A decision had not yet been made by him about the means by which the Department would increase the expertise. This would be the subject of proposals by Mr Petrie who had already made arrangements to devote part of his resources to the development of FSA. He would expect him to draw on the relevant expertise from the HSE. That was part of the integrated process which he was delighted to see in operation. In the context of expertise in the assessment of management systems he said: "I think there is certainly scope for increasing the expertise which exists but I would not wish to give the impression that there is no expertise." He had not yet seen Mr Petrie's proposals as to the new resources which should be introduced. He expected that in connection with the attracting of persons with the relevant expertise it would be necessary to supplement the training which the Department currently provided. As regards the timescale this should be related to the timescale within which FSA would be introduced. He thought it likely that the regulations would be made in the latter half of 1991. "The fact that the Department is not able yet to benefit from such things as formal safety assessment, which would go in greater depth into management systems, is something which we are working on, as you know, but I would not regard that as an indication that we are falling down on the job."

21.23   As regards the offshore inspectorate I have described in para 15.44 *et seq* the persistent shortfall in manning levels. Mr Petrie said that he had put in a bid to increase the total complement from 10 to 12 inspectors, which may be compared with 7 in post. In January 1990 this bid had not yet been put before the management board. At that time 9 specialists were being recruited. The divisional return which was to be submitted to the management board made provision for a further 8 specialists. The total of 17 would bring the total of specialist staff from 45 to 62. This took no account of administrators, including the safety policy branch where an increase in staff would help to take forward new regulations and assist generally with matters of policy. In the last few months there had been a "total rethink" of resources. A previous "fundamental rethink" in late 1986 had been implemented in 1987/88. Staff changes at that stage had given rise to many vacancies at the grade of Senior Inspector. Mr Priddle said, when giving evidence also in January 1990, that he was aware of the areas in which the Department had been unable to achieve because of lack of manpower. First attention had been given to the manning of the offshore inspectorate.

The HSE's approach to effective management of safety

21.24   It is of some interest to consider, by way of comparison, the approach adopted onshore. The HSE's approach to the management of safety as part of the Safety Case has been set out in Chapter 17. The HSC's plan of work for 1989-90 and beyond makes clear that the HSC and the HSE are intent on a vigorous promotion of the effective management of safety by industry. Leaving aside the premises to which the CIMAH regulations apply the HSE's approach is to insist on evidence that safety and particularly incidents are being properly monitored. The APAU, whom Mr Rimington described as "a sort of crack unit" carry out safety audits in co-operation with large companies and undertakings. Through this work they have acquired expertise in management systems and in packages for the monitoring of safety. They train other inspectors in management systems and increasingly take part in investigations of major incidents.

21.25   As regards inspectors' approach to onshore inspections reference may be made to what I have set out in para 16.37. Mr Rimington said that inspectors were trained particularly to concentrate on the management systems and attitudes in the course of their inspections. This was something for which they were most certainly trained. They could also call in the APAU if they had doubt.

The NPD's approach to the promotion of effective management of safety

21.26   Mr Ognedal said that safety could not be "inspected into a platform". Commenting on past experience of inspections he said: "In Norway it has had a tendency to create a situation where people do what they are told by these inspections and then wait more or less for the next inspection to come along and tell them what to do then." He elaborated this as follows:

   "We found that where we had identified a number of things on a platform requiring attention and had notified the operator of these, the operator would tend to react only to the matters drawn to his attention. We asked operators whether they were evaluating our comments on individual platforms across their platforms and fields and examining their systems in the light of the specific matters we were drawing to their attention. It appeared from the responses we received that this was not being done. We considered how we could focus on these issues with a view to motivating companies to do this themselves."

21.27   Mr Ognedal said that the main reason for starting to think along the lines of internal control was that NPD decided that the traditional way of supervising an activity was not effective. What was required was to get the operator to focus on the safety issue in a more systematic manner than was the case. He pointed out that even within the present system in Norway verification, including inspection, could only reveal what had gone wrong. This was the basic important point in auditing the procedures that controlled the activity. If weaknesses in these procedures were revealed, corrections could be made before they resulted in erroneous work performance or equipment failure.

21.28   He also expressed the view that it was a better use of resources to look at the framework that would produce safe activities than to find out only what had gone wrong. "I would say that my conviction is that my resources should be used to promote the operator and all his personnel, including contractors, etc, to be conscious in relation to the activities and the framework controlling the activities themselves, and to help and motivate that organisation which is present 24 hours a day. That is the best use of my resources."

21.29   As regards the implementation of internal control Mr Ognedal said that normally operators would base their method of management and control on accepted methods and standards for quality assurance. It had therefore been recognised that the duty of having an established internal control system was complied with by implementing integrated quality assurance systems, such as in accordance with ISO 9000. He preferred to see internal control integrated in this way.

21.30   Mr Ognedal said that 5 or 6 years ago consultants produced much of the documentation which was required for internal control. However operators had very quickly found out that they could not use documents which had been written by consultants. At the present time they produced the documents themselves and used consultants only to assist in defining what the scope and content of the document should be. Some operators had found the implementation of the system time-consuming. He was not sure if they found it difficult. It certainly took time to establish the system and go through the documentation which controlled the activity to see that it was coherent and that the system was implemented and properly understood in the organisation. His view was that without something similar in nature to internal control a company would be less safe. As regards contractors the operator would require to assess any contractor which it was going to use and check that there was some form of internal control activity within the contractor's organisation. The operator would also have a duty to audit that to see that it did what it was supposed to do and to correct any flaws that were found. While the licensee had a duty to see that there was a system of internal control the duty to participate in that system affected all who took part in work offshore in the petroleum industry.

21.31   Mr Ognedal agreed that in carrying out supervision the staff of NPD were now much more involved in making judgements; and had to have the ability not only to make them but to defend them in discussion with management. During the auditing process his staff were dealing much more with senior managers than was the case before. This meant that they required to understand the managerial role and the organisation that particular managers had under their control.

The allocation of responsibilities

21.32   The Burgoyne Committee recommended that the Government should discharge its responsibility for offshore safety through a single government agency whose task it was to set standards and to ensure their achievement (6.5). This was in distinction from the situation at the time of their report where 3 agencies namely the DEn, the HSE and the Department of Trade each had certain responsibilities for safety offshore. The committee envisaged that in the event of their recommendation being implemented the arrangement whereby the Department of Trade carried out examinations on an agency basis should be terminated and the DEn should assume that task (4.10). At the same time the committee firmly upheld what they called "the principle of independent certification of critical features of offshore structural and operational safety". They recorded that the practice of subjecting all aspects of the design and construction process to the independent scrutiny of a certifying authority had found general support and approval (4.25).

21.33   It is clear that the Burgoyne Committee did not perceive that there was any inconsistency between the concept of a single government agency and the role performed by a certifying authority, although the effect of that arrangement was that an important measure of discretion was entrusted to another body and that the process which led up to the issuing of the certificate of fitness was outside the direct knowledge of the government agency. It appears that the special expertise of the certifying authorities was seen as favouring this arrangement; and that their well-established standing provided a full assurance of independence. Since 1987 their work has been audited by the DEn. During the course of the Inquiry no criticism was made of their ability for, or their performance of, the work which has so far been entrusted to them. The attention of the Inquiry was drawn to the fact that they can undertake work as consultants in the design of installations. To carry out such a function would obviously be inconsistent with acting as a certifying authority in regard to the same matter. However, I am satisfied that in practice conflict would be avoided; and that the possibility of consultancy work should not affect the running of the certification system in any material way.

21.34 The DoT continues to act as the agent of the DEn in regard to fire-fighting equipment and life-saving appliances. Their work is also audited by the DEn.

21.35 During the course of the Inquiry a number of witnesses gave evidence that it was important that the work of regulation should be carried out by a single body. In Chapter 17 I have already discussed that point in relation to FSA. Mr M Ferrow, Manager, Safety and Quality Assurance for Conoco (UK) Ltd, spoke to UKOOA's position paper which advanced the view that it was "essential that the outside authority is competent in assessing both the engineering and management control aspects. Due to the integrated nature of the FSA there should be a single body responsible for the overall assessment." In that connection Mr Ferrow said: "We operators can direct our energies at safety and being safe more profitably by not being encumbered by a complex regime which requires us to interface with several bodies on specific matters. It would benefit everyone, in my belief, if operators could deal with one single authority who understood the overall issue at stake and could indeed help the operators to achieve their objectives."

21.36 Mr Ferrow also commented that the offshore safety regime had developed in such a way that offshore inspectorates were more fragmented than onshore. The pipelines inspectorate from his point of view appeared to be a relatively separate part of the DEn enforcing different sets of Acts and regulations. He disagreed that such separation was inevitable where special expertise was required. There was no single point of contact which looked at the overall issue.

21.37 In Chapter 19 I referred to the evidence of Mr Brandie, Safety and Compliance Manager, Chevron (UK) Ltd and the Chairman of the UKOOA Fire Protection Working Group, who maintained that the best technical solution to fire protection had been hampered by the splitting of the regulatory requirements for passive and active fire protection. Two sets of regulations were administered for the DEn by different authorities, namely the DoT and the certifying authorities. As a practical matter the latter tended to have a more continuous dialogue with operators. In connection with Chevron's own proposals it had been found necessary to have meetings with both the DoT and the DEn at the same time in order to make sure that they were not in conflict with the requirements of the Fire Fighting Equipment Regulations.

Regulations and guidance

21.38 After a review of the legislative and other controls exercised by the DEn for offshore safety the Burgoyne Committee came to certain conclusions as to the structure of the written controls. These were, *inter alia*, that an Act of Parliament "sets out main duties and obligations"; that regulations "detail mandatory objectives of controls"; and that guidance notes "relate to a set of regulations, give non-mandatory advice on methods of achieving objectives". (4.48). They recommended that: "Future regulations should specify objectives and avoid overlap. Methods of implementation should be advised as fully and flexibly as possible in guidance notes which should be recognised as being non-mandatory." (6.15).

21.39 From all sides during the Inquiry there was support for the proposition that the regime should be controlled by regulations which set objectives ("goal-setting regulations"). Mr Petrie agreed that where possible regulations should set objectives rather than lay down a series of prescriptive requirements, and that this was the way forward so far as the Department was concerned. This allowed for flexibility and the best practices to be used without inhibition. Goal-setting regulations were equally applicable in the area of mobile drilling platforms. The Department generally had the same approach as the HSE. He agreed completely that this placed a greater burden on the regulator in the sense that he must exercise his experience, judgement and discretion on areas which might be subject to debate. However, the distinction between a goal-setting regulation and a prescriptive regulation was very rarely clear-cut. Some existing regulations could be described as setting objectives. He did not wish to give up the tool of prescriptive regulation where it was appropriate in order to prescribe a minimum standard. Mr Priddle accepted that future regulations should in principle be goal-setting in their nature, but he observed: "Specific requirements seem to us very valuable in defining acceptable standards in certain well-defined cases." By way of comparison I may add that in Norway the NPD are currently moving away from a regime of detailed regulations and re-emphasising that the operator himself has to make the appropriate decisions based on objectives. Mr Ognedal stated that it was the intention of NPD to reduce prescriptive regulation to a minimum in all areas where it was possible to do so without affecting the safety level. It was foreseen that in some areas there would still have to be prescriptive regulations. One of the reasons for adopting the goal-setting approach was to make regulations that were more flexible, so that changing technology could be accommodated without the need for new legislation.

21.40 The movement towards goal-setting regulations would be in full accordance with the philosophy adopted by the Robens Committee for safety and health legislation. However, despite the statements of attitude made by witnesses from the DEn there has been virtually no progress towards the creation of new goal-setting regulations since the publication of the report of the Burgoyne Committee in 1980. Mr Petrie agreed that onshore the HSE had achieved this in certain areas. "We have not, as yet, managed that, although the target should have been reached." When asked for the reason for the lack of goal-setting regulations he said "I do not believe that our philosophy is that different from HSE. We have had difficulties with manpower to take forward this work, because of other work." Another factor was the task of considering the balance to be struck between goal-setting and prescriptive regulations. In the result the existing regulations under the MWA, most of which were made prior to the report of the Burgoyne Committee are different in their general approach from the type of goal-setting regulations which have been produced by the HSE on the basis of the HSWA.

21.41 Quite apart from this state of affairs it is clear that a number of the existing regulations under the MWA have already been recognised to be in need of up-dating. The PED's programme for 1988/89 stated that in view of the many changes in the offshore industry during the 10-15 years since many of the regulations had been made, it was intended to undertake a review of legislation with a view to up-dating, rationalising and streamlining regulations wherever possible. This was stated as one of the actions which the Safety Directorate intended to take with the overall objective of the improvement of safety standards offshore and the corresponding reduction in accidents. Mr Petrie explained that this project had reached the point of identifying priority areas. However, due to constraint on resources that existed even before the disaster it had been agreed that it was no longer feasible to consolidate and streamline the regulations in the way originally envisaged, as the programme for the following year makes clear. However that programme indicated a proposal to prepare a timetable for (i) the Operational Safety, Health and Welfare Regulations, which "are far too specific with the result that they have become out of date. Consequently a large number of requests for exemption from some detailed requirements of the regulations are received"; (compare para 7 of the DEn's submission to the Burgoyne Committee in 1979); (ii) the Life-saving Appliances Regulations and the Fire Fighting Equipment Regulations, where "technological changes in recent years mean that some up-dating is necessary"; and (iii) other regulations including the Emergency Procedures Regulations, which needed to address the question of one standby vessel supporting more than one installation.

21.42 During the course of his evidence Mr Ferrow advocated the approach that: "Efforts should be strongly directed at safety rather than compliance for its own sake." He explained: "It can be extremely expensive and disruptive to carry out certain specific precautions and buy very, very little or, in fact, even negative value in terms of safety, simply in order to comply with a regulation." He went on to say: "The problem with the regulations as they have existed is that they do not address the overall system whereby the individual components are connected together, so, whereas there are very particular design codes, etc for valves and pressure vessels and so on, the way in which all those particular components interact is in fact not the subject of any particular specific legislation that I am aware of for offshore platforms at the moment." The point made by Mr Ferrow is similar to one made by ICI Petroleum Services Ltd in their submission to the Burgoyne Committee which is printed at pages 239-240 of their report. They stated in the course of their submission:

"Experience onshore since the introduction of the HSWA compared with the previous legislation seems to be that the principles of self-regulation and management control are resulting in a more responsible forward-looking attitude by companies. The present system of control by regulation in the North Sea could lead, it is believed, to an attitude on the part of some employers whereby there is a primary desire to comply with the regulations rather than exert maximum effort towards total safety. Moreover, regulations are slow to form and difficult to change; they are inappropriate for complex and rapidly changing technologies, and they are capable of being abused by encouraging the attitude typified by 'the plant must be safe because everything has been done that the regulations require'. What is needed for future projects is a more flexible system which can not only respond quickly to new problems - thereby generating improvements - but encourage a forward-looking attitude and put the initial responsibility for deciding what is safe where it belongs - with the employers."

At this point it is worth recalling the quotation from the evidence of Mr McKee which I gave in para 21.4, to which may be added the following quotation from his evidence: "Regulations need to be less prescriptive and detailed, more objective and broader based. Over time as you layer more and more prescriptive types of regulations on to the overall regime it probably takes away from the overall objective of total safety." By way of comparison I noted that Mr Rimington's evidence was that regulations under the HSWA normally address themselves to systems. He cited in that connection the provisions of the Pressure Systems and Transportable Gas Containers Regulations 1989 under which management were required to validate and confirm pressure systems in a systematic way. He went on to say: "Now accompanying those regulations, which principally address themselves to systems and responsibilities, are quite considerable codes that will go into all sorts of details, such as, for example, what you do if you come to a pressurised system which you can enter, as opposed to one that you cannot, and so on. Therefore, we proceed from the general to the particular, leaving varying degrees of latitude to an employer as to how he tackles the particular. That is our whole philosophy."

21.43 Mr Ferrow also put in a plea that the structure of Acts, regulations and guidance should, if possible, be made simpler. He said that there would be great benefit in a system of legislation which all, including engineers, operators as well as managers, would understand. He saw the potential for a simpler framework of legislation that did not remove previous legislation at a stroke but looked towards simplifying and incorporating it within that framework. He thought that the better way to go was to set up a requirement for assessments rather than attempting to identify all the particular hazards that there could be in all situations and then providing particular rules to address those particular points.

21.44 Mr Ferrow went on to say that it was an unsatisfactory situation where it was necessary to apply for exemptions to strict regulations. He would be happier dealing with guidance which could be discussed on a case by case basis. This point was supported by other witnesses who were concerned with various measures for safeguarding personnel in the event of an emergency. Thus Mr M Booth, Head of Operations Safety, Shell (UK) Exploration and Production Ltd who dealt with escape routes said that in view of the diversity of platforms it was undesirable to have a detailed prescriptive legislative approach. It stifled technology and advancement and at the end of the day it was counter-productive. A similar point was made by Mr I Wallace, Superintendent of Occupational Safety and Health, Conoco (UK) Ltd, and Dr J Side of the Institute of Offshore Engineering, Heriot-Watt University, who dealt with emergency evacuation, escape and rescue. The dangers of over-prescriptive regulation are I think clear from this evidence. It is unwise for any regulator to put much reliance on exemptions which take time and trouble to obtain and may discourage an operator from incorporating the benefit of improvements in technology. Further, as Mr Petrie accepted, the fact that any regulation requires a large number of exemptions may well indicate that the regulation has been badly framed.

21.45 As I stated above in para 21.38, the Burgoyne Committee recommended that guidance notes should be recognised as being non-mandatory. They further said that: "The guidance notes to regulations should be kept up to date on a continuous basis and their status as non-mandatory guidance should be clear." (6.17). Mr Petrie said: "There is frequently a misconception on the part of people who do use guidance notes that they are more than guidance notes. It will not be the first time that somebody has said to me they wanted to talk about regulations when in fact they were talking about guidance notes. We always point out that it is exactly guidance notes and the standing of them. Of course, it is explained in the front of the guidance notes that they are non-mandatory." On the other hand the PED saw guidance notes mainly as giving minimum standards. He did not agree that this represented the application of guidance notes in a prescriptive manner, because the same level of safety might be achieved in another way. "When I said minimum I mean meeting the guidance notes provides the minimum standard that we believe will comply with the regulations. There are other ways of achieving exactly the same thing - or potentially other ways of achieving the same thing. That is why the guidance notes are not mandatory to allow that flexibility." However, a number of witnesses spoke of the problems created by the fact that guidance notes tended to be treated as an obstacle to alternatives. Mr Ferrow said of the guidance notes which related to the Construction and Survey Regulations: "I can assure you that they are adhered to almost to the letter and are taken extremely seriously by engineers at all levels. It has become a working document for the method of construction and principally structural matters, and now extends into a wider variety of matters. As a guidance document it works almost too well in the sense that the certifying authorities seem unprepared to deviate from the written guidance without reference back to the DEn." Referring to the guidance notes relating to Reg 11(1) of the Fire Fighting Equipment Regulations he said: "The requirement to provide fire-water in process areas at the rate of 12.2 litres/m²/minute is, in my view, off the point of providing fire-water to mitigate the consequences of a fire. If I think of the resources that are brought to bear in terms of trying to achieve those sort of objectives, I feel they would be better employed looking at and analysing what the likely fires would be and what the necessary rates of fire-water would be for that particular incident." cf para 19.66 for the evidence of Mr Brandie on this matter.

21.46 I turn now to examine the history of the extent to which regulations made under the HSWA have been extended to offshore as well as onshore application. The Burgoyne Committee noted that safety offshore was the subject of the MWA, the HSWA and the PSPA and regulations thereunder. They saw as the first task of the single agency for safety offshore to review the overlap between these Acts (4.24). They recommended that future regulations should avoid overlap (6.15). They commented that the further development of safety regulations could in theory be undertaken under either the MWA and the PSPA on the one hand or the HWSA on the other, although the wider application of the latter made it preferable (4.21). The agency agreement which was entered into by the HSC and the Secretary of State for Energy set out that it had been agreed between them that the Secretary of State would make adequate arrangements in accordance with such guidance as the HSC or the HSE might give for the development of health and safety regulations, approved codes of practice and other advisory material under the HSWA. The revised letter of implementation dated 23 March 1982 reflected this. The HSE seconded inspectors to the PED in the expectation that they would put forward regulations under the HSWA. There appears to have been no problem in framing regulations under the HSWA in such a way as to place duties on licensees and other specific categories of persons defined in the MWA. In the event no legislation under the HSWA has been promoted by the PED apart from the Offshore Installations and Pipeline Works (First Aid) Regulations 1989. Further, in the light of the attitude and advice on policy given by the PED the offshore application of sets of regulations prepared by the HSE in the modern form which is in line with the views of the Robens Committee and the policy of the HSWA has occurred only in a limited number of cases. It will also be recalled that whereas since 1977 the HSWA itself has applied to the UKCS, the legislation relating to the offshore industry such as the MWA and its regulations have not been made relevant legislation for the purposes of the HSWA and accordingly subject to replacement under that Act.

21.47 Mr Rimington explained that the procedure when the question of offshore application arose was that the HSE's policy branch asked the PED what advice it would give to the HSC. The policy on offshore application was not a matter in which the HSE played any direct part. This was done under the agency agreement. At a later stage it became a policy decision on the part of the HSC whether or not to accept whatever advice they received from the PED. Out of 27 occasions in which the PED's advice had been solicited the answer on 7 occasions was in favour of, and on 20 occasions was against, offshore application. In regard to those figures Mr Rimington observed: "That is a very limited way of perhaps looking at the matter. I have to repeat that I am not very familiar with the offshore situation. I also have to say that the Mineral Workings Offshore Acts are, though perhaps old-fashioned in form, really quite a modern set of provisions. They in general seem to have served the industry well in the estimation certainly of those who are regulated and also, in the view of the DEn, who are much closer to the matter than we are." The regulations in the case of which the PED advised against offshore application included proposals as to CIMAH Regulations in 1983; and Control of Substances Hazardous to Health COSHH Regulations in 1985. In Chapter 22 I will have some comments to make in regard to the PED's understanding of and attitude to the offshore application of those proposals. Mr Rimington said that it was clear from these and other examples that the response from the PED was identical. Since clearly some general factor was at work the HSC invited the PED to produce a view on the relative operation in the future of the HSWA and the MWA "because clearly there was a tension between the two". He explained that the overlap between the Acts was of the liveliest concern to him and, he believed, the HSC. As the HSWA regulations extended offshore, if indeed that was intended, the position of the HSE under the agency agreement automatically extended. "Given the fact that the HSWA in any case applies offshore, and is indeed used by the PED, very considerable difficulty arises in knowing precisely what the extent of the executive's responsibility is, or indeed the commission's responsibility ... so the commission wished to have the policy of the DEn set out much more clearly that it had been. Such a policy, when set out, could not obliterate the difficulty of 2 overlapping sets of legislation, but it would produce more stability in the situation." He also said that the HSE had always perceived in its relationship with the PED a very great sensitivity which he personally could understand, about the question of policy for anything that had to do with the structural integrity of offshore installations. He said that the PED "had a very important and comprehensive system going. It had been endorsed by the major committee that had sat on the subject. We took it that this was a very considerable bar to them feeling that some very substantial change involving legislation should be undertaken."

21.48 In 1986 the PED submitted a written review of offshore safety legislation in response to the request from HSC. It stated that there appeared to be no problems relating to the interaction between the PPA, the MWA, the HSWA and the PSPA as they affected health and safety. New regulations under the HSWA could be used as an "infill" to existing requirements, sometimes as an "addition" and on occasions as in "substitution" for existing standards. On the other hand it said that, however welcome such assimilations might be, this general rule could not be applied universally. Offshore there was a need to apply specific standards which might have no logical counterpart onshore, such as in the area of first-aid, emergency procedures, fire-fighting and life-saving. Further, certification had no direct onshore counterpart. Its efficacy had withstood the test of time and there was a clear need to continue with this concept. The review went on to say:

> "Whilst arguments could be advanced to make the 1971 and 1975 Acts relevant statutory provisions of the 1974 Act, it is believed that the present position is essential for current and foreseeable offshore needs. There are 2 overriding and salient features of mineral workings legislation: that of 'structural integrity' and that of exploration, including drilling and production licences and consents - matters which are not in themselves dependent on or relevant to the occupational health and safety of persons or directly affecting the public. It is self-evident that there is a need to ensure such standards whether persons are employed or not. Unmanned installations are likely to increase as is the number of sub-sea completion systems. There will therefore be a need for regulations under the 1971 and 1975 Acts to ensure the continuation and improvement of appropriate standards. With offshore safety policy clearly placed in the hands of the Secretary of State for Energy there would appear to be no conceivable reason to divest him of this responsibility as a paper exercise only to return it to him as part of a new agency agreement since the existing agreement and letter of implementation exhort him to develop health and safety legislation by utilising the HSWA. This paper acknowledges and accepts that occupational health and safety should, where ever appropriate, be made under the HSWA. The Offshore First-Aid Regulations are a prime example of this action and shows the co-operation between the executive in this department in their development as well as the acceptance of 5 other sets of HSWA regulations which have an offshore application."

Among the conclusions of the paper it was said that there was a need to retain specific regulations in such matters as life-saving, fire-fighting and emergency procedures. "Such regulations could be made under the HSWA: although historically having no onshore connotation it would seem logical to continue any future action under the 1971 and 1975 Acts for the sake of continuity."

21.49 Mr Rimington commented that the paper very largely confirmed his under-standing as to the view which the PED took as to the scope for the use of regulations under the HSWA. "What it does not resolve is - given that regulation of anything has to be tackled from a certain philosophy - any conflict between the philosophies involved. In order to give a final answer to the question, how far are we entitled to form judgements of PED's effectiveness? Where does that judgement begin or end? Does it stop entirely with the 7 regulations, or does it extend somehow beyond that? If it extends beyond that, how do you cut them off?" Mr Rimington also added that the scope of the expression "occupational health and safety" was extremely difficult to define. However, he accepted that the paper achieved what the Burgoyne Committee had referred to (in 4.24) as the review of the overlapping Safety Acts. The PED had streamlined the situation in the sense of relying almost wholly on the MWA. In those circumstances he thought that they could very fairly claim to have simplified the situation and brought some order into it.

21.50 Mr Petrie explained the approach which the PED took when presented with proposed regulations. He said: "We look to see the purpose and intention of any proposed regulations, see how they are perhaps already covered by our existing regime. If they are not considered to be already covered or partially covered, then the decision has to be taken as to whether it is appropriate for that particular bit of legislation to apply offshore." As a matter of general policy he would not necessarily be against disturbing "the old regime". This had been done in a number of cases and he expected to do that on occasions. "But I think perhaps I should also make it clear that we assess the situation and put the case to the HSC, who are, if I can so call it, the final arbiter in that matter and advise the Secretary of State of their view after having heard evidence, information, views of the department." He added later: "I think it is a general view that we should not unnecessarily be amending and changing legislation. Indeed, in that same light, there are many onshore regulations that are still not assimilated in new regulations under the 1974 Act. Assimilation would certainly be striven for in the case of new regulations which have no existing counterpart in the offshore safety regime. But even if existing regulations were merely being amended the PED would look into the possibility of assimilation." Mr Priddle repeated the point that the HSC had the right to advise the Secretary of State of a contrary view from that expressed by the PED.

21.51   My general conclusions in the light of this evidence are that there has been virtually no progress towards the creation of new goal-setting regulations. Many existing regulations are unduly restrictive in that they are of the type which impose 'solutions' rather than 'objectives'; and are out of date in relation to technological advances. Guidance notes are expressed, or at any rate lend themselves to interpretation, in such a way as to discourage alternatives. This poses a clear danger that compliance takes precedence over wider safety considerations; and that sound innovations are discouraged. The PED have advised a policy - which the HSC has accepted - of reliance mainly on the MWA. Even if it is accepted that structural integrity is a special feature of what is required in the case of offshore installations and that it should not be forgotten that unmanned installations form part of the total number, there is a clear overlap between the field covered by the HSWA and that covered by Section 6 of the MWA. That section forms a basis for a number of the existing sets of regulations, including the Operational Safety Health and Welfare, Emergency Procedures, Life-saving Appliances and Fire Fighting Equipment Regulations. In the result, on the one hand the existing regulations under the MWA have stagnated; and on the other hand the effect of the policy advised by the PED has been to distance offshore regulations from the influence of the main stream of practice in modern regulations on health and safety. One outstanding example of the result of this policy was the rejection in 1983 of the offshore application of the proposed CIMAH Regulations. The same point applies whether one talks of the extension offshore of regulations which are to apply onshore or the creation of a parallel set of regulations which are adapted to conditions peculiar to offshore installations.

The future regulatory approach

21.52   In this part of the chapter I shall draw together a number of matters which have been discussed earlier and set out what I consider to be the changes which are required in the regime in addition to those already recommended on the basis of Chapters 17-20.

Ensuring the adequate management of safety by all operators

21.53   The earlier discussion shows that although operators' management of safety is of critical importance to the safety of personnel on installations the present regulatory provisions do not address it in any direct sense. They consist on the one hand of sets of regulations which are directed to specific and limited subjects (mainly 'hardware') and on the other hand the broad duties set out in the HSWA. Further, I am not convinced that under the present regime the regulatory body can monitor and support the operators' management of safety in more than a minor and incidental way. My views would remain unchanged even if the Safety Directorate were fully manned - which is far from being the case at present.

21.54   It is also clear to me that the offshore safety regime has fallen significantly behind the onshore regime in a number of respects in which thinking on safety matters has advanced over the last 10 years. The respect which is most relevant at this point is the concept of the Safety Case. The DEn's advice against not only the offshore application of the proposed CIMAH Regulations but also the facility of a Safety Case has set back the development of the offshore safety regime by many years. Even though the Safety Case has proved to be more successful than could have been predicted by the HSE its introduction was rightly regarded as a major advance in the technique of the regulation of safety.

21.55   Accordingly in my view a number of major changes in the offshore safety regime are long since due. These have implications both for operators and for the regulatory body. In previous chapters I have recommended the introduction of a requirement for submission of a Safety Case for various purposes which take account of the peculiar problems presented by offshore installations. In my view it is necessary to go one stage further in order to ensure that operators set out their system for the management of safety and demonstrate that it is adhered to.

21.56   I consider that operators should be required to set out formally the safety management system which they have instituted for their companies and to demonstrate that it is adequate for the purpose of ensuring that the design and operation of their installations and equipment are safe. For convenience of reference in this report I will refer to the safety management system as SMS. The SMS would be expected to set out the safety objectives of the operators, the system by which those objectives were to be achieved, the performance standards which were to be met and the means by which adherence to those standards was to be monitored. The SMS would be expected to contain a full demonstration as to how safety was to be achieved in both design and operation. Thus it would cover, *inter alia*, how safety was to be achieved through:-

— organisational structure
— management personnel standards
— training, for operations and emergencies
— safety assessment
— design procedures
— procedures, for operations, maintenance, modifications and emergencies
— management of safety by contractors in respect of their work
— the involvement of the workforce (operators' and contractors') in safety
— accident and incident reporting, investigation and follow-up
— monitoring and auditing of the operation of the system
— systematic re-appraisal of the system in the light of the experience of the operator and industry.

21.57   It would be appropriate that this demonstration should form a leading part of the Safety Case, for which much of the information would be required in any event in connection with major hazards. Along with the SMS would be any safety management system which was particular to the installation for which the Safety Case was prepared.

21.58   I have considered carefully whether or not I should recommend that the SMS should be set up in accordance with any particular type of system which is already in use. At the Inquiry Dr A A Denton, giving evidence on behalf of the Institution of Mechanical Engineers, advanced the view that operators should be required to adopt quality management systems (QMS) techniques. This would involve the application of QMS to the whole of a company's operations of which the management of safety formed part. Dr Denton defined quality management as all systematic actions which were necessary to ensure that the activity is planned, organised, executed and maintained according to requirements in, and pursuant to, laws and regulations, and in adherence to corporate policies, requirements and specifications. QMS control what must be done; who will do it; how it will be done; if it must be controlled by instructions, procedures or drawings; how the accomplishment of the task is to be documented; who will verify that the work was completed as planned; and what records must be kept, by whom, and for how long. QMS had 4 "prime indispensable and indivisable components", namely a corporate quality manual and subsidiary quality manuals for individual platforms; a requirement that the manual be followed; regular audits by an independent third party; and a response to deficiencies by appropriate corrective action. Dr Denton maintained that total quality could not be applied to safety alone. Hence if QMS were required by regulation to apply to safety it would force a company to apply QMS to every part of its activities. I have come to the conclusion that it would be going too far for me to recommend the imposition of a system which would apply to all operators and across the entirety of their operations. I take the view that the operators should have the freedom to choose the type of system which is appropriate for them, in the light of the regime's requirements and their own operations. However, in the light of the evidence which I have heard I consider that in the formulation of their SMS operators should draw on principles of quality assurance similar to those contained in BS 5750 and ISO 9000.

21.59   I should perhaps add that as part of his evidence Dr Denton proposed on behalf of the Institution of Mechanical Engineers that mandatory minimum standards of technical qualifications should be established for platform staff. cf para 20.58. For example, while accepting that the exact qualifications would depend on the size and complexity of a platform, the OIM should be technically literate at least to Higher National Diploma or equivalent standard and have at least 3 years offshore experience and an understanding of the sea/air environment. An operations superintendent should be a Chartered Engineer, process operators should be qualified to Higher National Certificate or equivalent standard, and each specialist maintenance trade should be led by someone with at least Higher National Diploma status. In any managing position the occupant required both a sound theoretical understanding and relevant practical experience, preferably offshore. UKOOA submitted that it was for the operator to decide the appropriate manning levels for an installation and the appropriate qualifications of personnel. Technical qualifications needed to be balanced against other desired capabilities, such as skills in man management and communications abilities. I am not persuaded that specifying standards of technical education for the generality of platform positions is a practical way forward, as platforms vary in size and complexity, as do the organisational systems of operators. However the competence, including the soundness of technical understanding, of those appointed to positions of authority is an issue critical to the safe operation of any platform and, while agreeing that this has to be for the decision of the operator, it should be set out for review by the regulator as part of the operator's SMS.

21.60   It is clearly essential that in addition there should be controls by means of which the regulatory body can be assured that the SMS is adhered to. It is clearly inappropriate and impracticable for the regulatory body to be made responsible for auditing in detail operators' compliance with their SMS. Accordingly, it should be part of the regime that operators are required to satisfy themselves by means of regular audits that their SMS are being adhered to. On the other hand the regulatory body should be required regularly to review operators' audits on a selective basis; and itself to carry out such further audits as it thinks fit; and by regular inspection verify that the output of the SMS is satisfactory.

21.61   What I have outlined in the last paragraphs involves a completely new approach to regulation in the UKCS. It is, however, totally consistent with the HSWA and the concept of self-regulation. It represents in my view a logical development from the requirement of a Safety Case for each installation. It is true that it has no current counterpart onshore. However, it can be seen as a further advance in the philosophy of a safety regime. Further, the evidence has shown that the industry consists of a relatively small number of companies running high technology operations where there is a strong need for a systematic approach to the management of safety. In any event its introduction offshore could have ultimate benefits for the onshore safety regime. The statutory assessment of the management of safety by the use of SMS offshore parallels the work of the APAU which is undertaken by agreement with employers onshore. In the light of evidence as to what operators are already accustomed to do in the UKCS and the NCS I am confident that operators will be able to adapt to this change in the regime. As regards the regulatory body, these and other changes will call for expertise and resources well beyond those presently enjoyed by the DEn.

The allocation of responsibilities under the regime

21.62   I am entirely satisfied that I should endorse the view which the Burgoyne Committee expressed that there should be a single regulatory body. (cf para 17.71.) While even within a single body there are inevitably separations due to differences in expertise and function there are clearly advantages in the co-ordination of the work of regulation. This is particularly important for the future in which a greater burden will be placed on the expertise, judgement and resources of the regulator, upon which his confidence and that of the industry will rely.

21.63   It is clear to me that, given the introduction of a requirement for a Safety Case, and the associated requirements for operators to demonstrate their SMS and audit compliance with it, the need for certifying authorities to continue to perform the same functions as before should be re-appraised. A number of parties to the Inquiry submitted on the basis of the introduction of FSA or 'internal control' that the present role of the certifying authorities should be brought to an end. For example UKOOA supported the submission that the certificate of fitness should in future be granted by the regulatory body on the basis of a survey and report by one of the existing certifying authorities or any other satisfactory body, subject to the inclusion of the operators themselves. The Contractors' Interest on the other hand submitted that certifying authorities should continue to be responsible for examination of 'hardware'; whereas the assessment of management systems should be the responsibility of the regulatory body.

21.64   Having considered those submissions in the light of the evidence I have come to the conclusion that it would be going too far and too fast for me to recommend particular changes in this area. I consider that it is not advisable or practicable for me to make a re-appraisal. This is a matter which should be carried out by the regulatory body. The other changes in the regime which I am recommending are in themselves major and will require a substantial amount of time and resources to plan, organise and implement. Their exact formulation is beyond the scope of a public inquiry. At present it is impossible to foresee all the considerations which may be of relevance and importance at these future stages. In these circumstances I consider that my best course is to recommend that the regulatory body should consider (i) after the introduction of requirements for demonstration of SMS and auditing of compliance with it; and (ii) after experience in the operation and effectiveness of such requirements whether and to what extent it will be appropriate to retain the present system of certification.

21.65   It remains for me to consider the position of the DoT. As will be seen from Chapter 19 I am strongly of the view that an integrated approach should be taken to fire protection so that both active and passive are considered together. To some extent this will be achieved through the Safety Case. I have, however, recommended that new regulations and guidance notes should promote such an integrated approach. In these circumstances it will be even more inappropriate than it is at present that different bodies should be concerned with separate consideration of active and passive fire protection. The ideal solution would be if these matters were considered wholly by the single regulatory body itself. This is of course complicated by the existence of the certification system. However as a first step I would advise that the regulatory body should assume direct responsibility for the functions which are presently discharged by the DoT. Further, I cannot see any sound reason for not adopting the same approach in regard to life-saving appliances.

21.66   As I am strongly in favour of a single regulatory body I consider that that body should discharge the regulatory functions in regard to standby vessels whether directly or through the agency of the DoT, save those which relate to the statutory responsibility of the DoT under the Merchant Shipping Acts.

Regulations and guidance

21.67   I am entirely satisfied that the principal regulations in regard to offshore safety should take the form of requiring that stated objectives are to be met rather than prescribing that detailed measures are to be taken. In relation to such regulations guidance notes should give non-mandatory advice on one or more methods of achieving such objectives without prescribing any particular method as a minimum or as the measure to be taken in default of an acceptable alternative. On these points I endorse the recommendations of the Burgoyne Committee at 6.15 and 6.17. However, I accept that there will be a continuing need for some regulations which prescribe detailed measures.

21.68   In connection with the proper development of offshore regulations it is in my view appropriate and necessary that the parts of the MWA and PSPA which have the same general purposes as those of Part 1 of the HSWA and any regulations made under those provisions should be made relevant statutory provisions for the purposes of the HSWA. The exact identification of the provisions in question is a matter which should be left to the regulatory body.

21.69   The replacement of the present sets of regulations with goal-setting regulations will obviously take some considerable time to execute. The regulatory body will have to decide what place this should occupy in the order of priorities, having regard to other major changes. There is clearly room for rationalisation of regulations, particularly having regard to the shape of the future regime. With those considerations in mind I consider that an appropriate form of replacement for the Construction and Survey Regulations, the Fire Fighting Equipment Regulations, the Life-saving Appliances Regulations and the Emergency Procedures Regulations would be:-

(i)   Construction Regulations, covering *inter alia* the structure and layout of the installation and its accommodation.

(ii)   Plant and Equipment Regulations, covering *inter alia* plant and equipment on the installation and in particular those handling hydrocarbons.

(iii) Fire and Explosion Protection Regulations, covering *inter alia* both active and passive fire protection and explosion protection, and

(iv)   Evacuation, Escape and Rescue Regulations, covering *inter alia* emergency procedures, life-saving appliances, evacuation, escape and rescue.

The text of Chapters 19 and 20 provides a number of examples of regulations which it would be appropriate to incorporate in these sets of regulations.

21.70   Operators should be encouraged to specify standards to be used by the company with a view to demonstrating compliance with goal-setting regulations. Thus in the case of a given installation operators may demonstrate compliance by reference to such standards, the terms of guidance notes and what is shown by a safety assessment or a combination of one or more of such methods.

21.71   As regards existing guidance notes the regulatory body should consider whether and to what extent they should be treated without replacement or modification as giving non-mandatory advice in the sense set out in para 21.67; and should inform the industry accordingly.

21.72   In the light of representations made at the Inquiry by the contractors' interests I would also advise that in connection with the preparation of guidance notes the regulatory body should review the procedures for consultation so as to ensure that the views of representatives of employers and employees involved in work offshore are adequately taken into account.

Involvement of the workforce

21.73   In para 18.48 I referred to the involvement of the workforce as an important means of developing and maintaining an attitude to safety which is conducive to the prevention of accidents which may have harmful consequences. In para 21.56 I indicated that the operators' SMS, which is directed to demonstrating how safety is to be achieved, should include the way in which the total workforce is involved to that end.

21.74   Under the present regime, both onshore and offshore, specific requirements have been laid down for the appointment and functions of safety representatives of the workforce. At the Inquiry there was a clear controversy, which I will deal with below, as to the form which the requirements in the offshore safety regime should take. However, the need for such requirements, whatever form they take, would not, in my view, be affected by the implementation of the recommendations which I have made so far in this report. The representation of the workforce in regard to safety matters is important not merely for what it achieves on installations but also for the effect which it has on the morale of the workforce - in showing that their views are taken into account and that they are making a worthwhile contribution to their own safety. For this purpose it is clearly advisable to have statutory provisions which are well known, universally applied in similar circumstances and effective in operation.

Safety representatives and safety committees in the onshore safety regime

21.75   Under Sec 2 of the HSWA regulations may provide for the appointment by "recognised trade unions" of safety representatives whom the employer is bound to consult in regard to arrangements for co-operation in the promotion and development of measures to ensure health and safety at work and in the checking of the effectiveness of such measures. The employer may be required to establish a safety committee which has the function of keeping under review the measures taken to ensure the health and safety at work of his employees and such other functions as may be prescribed. So far as the onshore safety regime is concerned these provisions were implemented by the making of the Safety Representatives and Safety Committees Regulations 1977, which confer various functions on safety representatives including the making of investigations, inspections and representations. A "recognised trade union", which had the sole power to appoint safety representatives, meant an independent trade union which the employer concerned recognised for the purposes of negotiations relating to or connected with one or more of a number of specified matters - such as the terms and conditions of employment, or the physical conditions in which any workers are required to work; the allocation of work or the duties of employment as between workers or groups of workers; and facilities for officials of trade unions. Since 1977 there has been a growth in the extent to which trade unions have been "recognised". Mr Rimington said that safety representatives could play a valuable part in the promotion of safety and in relation to inspections. For those who were appointed safety representatives it was a very great strength that they were appointed by the unions. "The unions train them in quite a sophisticated way. They have the means of putting a great deal of power at the elbow of safety representatives where they care to do so." Where a union was weakly organised or not very strongly represented the usefulness of the safety representatives might be somewhat impaired.

Safety representatives and safety committees in the offshore safety regime

21.76   Although Sec 2 of the HSWA applied to the UKCS from 1977 the 1977 Regulations were not applied offshore. Diametrically opposed views were held by the trade unions and UKOOA. The latter objected to the offshore application of the 1977 Regulations on the ground that there were very few installations where there was a "recognised trade union". The Burgoyne Committee supported the view that on each installation there should be a safety committee which was representative of the workforce, including contractors' personnel, but did not consider it essential to embody this in regulations (6.50 and 5.97). However, 2 members of the committee, Mr R Lyons, then National Officer of ASTMS and Mr J Miller, then National Officer of the T & GWU dissented strongly on the latter point, urging that the 1977 Regulations be extended offshore forthwith.

21.77   In the event after years of discussion the DEn in 1987 were able to achieve a measure of general acceptance which led to the making of the Offshore Installations (Safety Representatives and Safety Committees) Regulations 1989. These were made under the provisions of the MWA and provide that the workforce is to be entitled to elect safety representatives and that where these have been elected a safety committee is to be established. This was clearly a step forward and an attempt to deal with a real problem. It still left as the bone of contention whether safety representatives should be appointed by trade unions, as was the case onshore.

The trade unions' evidence

21.78   The attitude of the trade unions on the matter of safety representatives was one of the principal subjects of the evidence given by Mr Lyons, since 1987 the Assistant General Secretary of ASTMS and latterly of MSF; Mr F Higgs, National Secretary of the Chemical, Oil and Rubber Group, T & GWU; and Mr A W T Cunningham, Occupational Health and Safety Officer, EETPU. Mr Lyons said that MSF had over 4000 paying members and represented in total about 6000 employees in North Sea activities. MSF members worked for both operators and contractors and performed a variety of jobs. According to the evidence of Mr Higgs T & GWU had about 3000 members offshore.

21.79   It was clear that the background to the evidence of these witnesses was a long-standing frustration as to the limited extent to which trade unions had been "recognised" offshore; whereas the unions had been recognised by many of the operating companies in relation to their operations onshore. As Mr Lyons put it: "There is a large trade union influence offshore. It has not got an adequate machinery through which it can be expressed." He complained that a memorandum of understanding as to the procedure for achieving recognition had not been adhered to or enforced. There were members of MSF on every platform in the North Sea, he thought; and there was a majority membership of MSF alone on quite a few of the platforms where no ballots as to recognition had been agreed. "In many of the platforms we have got 100% membership." The Inter-Union Off-shore Oil Committee (IUOOC) had been formed in order to eliminate inter-union disputes over representation offshore. On behalf of the IUOOC he had entered into a recognition agreement in 1978 with the Phillips Petroleum Company in regard to platforms in the Hewett field, the effect of which was that the 1977 Regulations should be treated as if they applied offshore. He said that this had led to an improvement in practices and an increase in confidence. It was hoped to extend that agreement to the Maureen field. MSF had also made many agreements with Shell on behalf of Shell Exxon which were supported by ballots of the workforce. MSF was the only trade union which held such agreements. However in each instance the agreement excluded health and safety. He claimed that there was no other country in the world in which there was a practice whereby a trade union which had been recognised by the employer was excluded from discussing health and safety.

21.80   Mr Lyons castigated the 1989 Regulations as contrary to the spirit of the HSWA. Without the offshore extension of the 1977 Regulations it was nonsense to say that the HSWA fully applied offshore. The 1977 Regulations had the advantage that safety representatives appointed by trade unions would have the back-up and facilities which a trade union is able to provide, including training and advice on health and safety issues. Unions held regular training schools at which a wide range of health and safety issues were discussed. These took place at regional, national and international levels. If a safety representative had difficulty in performing his or her function there was somebody for him or her to go to in order to get assistance. "For a safety representative to be effective he requires a supportive culture, structure, credibility, advice, training and recognition of the contribution that he can make on safety issues." For a number of years Shell had had a safety committee system which was similar to that provided for under the 1989 Regulations. However, despite the efforts of MSF the workforce were reluctant to stand or be represented. Where the trade union appointed the safety representatives "training and advice can be given openly without any 'fear factor' which unfortunately permeates the UK sector of the North Sea among the workforce. Workers do not want to put their continued employment in jeopardy through raising a safety issue that might be seen as embarrassing to management." As an example he said that contractors' employees suffer particularly from the "not-required-back" phenomenon. When asked whether a safety committee elected by the whole workforce might be seen to be more representative than one which was restricted to members of trade unions he said: "The quality of that committee bears no relationship to a trade union-based safety committee, and that is best borne out by looking at Shell onshore, where the committees do not cover all employees but are extremely positive in health and safety." The 1989 Regulations were perceived as favouring the operators. This was seen as part of the evidence of a conflict of interest which led to trade unions favouring the replacement of the DEn with the HSE as the regulatory body, as he and Mr Miller had also advocated in their dissent from the report of the Burgoyne Committee.

The submissions of UKOOA

21.81 UKOOA opposed the application offshore of the 1977 Regulations. It would have only a limited scope for operation in view of the limited extent to which there were "recognised trade unions". The 1989 Regulations were adequate. They did not prevent a trade union member becoming a safety representative and having trade union support. There was no suggestion that trade union members were more concerned than others with matters of safety. Where trade unions represented a minority of the workforce, if they were able to appoint the safety representatives they might effectively disenfranchise non-union members: or even union members who might wish to have a different representative.

Safety delegates in the Norwegian offshore safety regime

21.82 In this regime it appears that trade unions receive automatic recognition. The extent of union membership has grown over the years. The regime provides for the appointment of safety delegates upon whom a number of important powers are conferred, including the right to halt dangerous work. Mr Ognedal considered that union back-up could be beneficial to the work of safety delegates. However, they are elected by the whole workforce, rather than being appointed by the unions.

Observations

21.83 My remit does not extend to matters of industrial relations, whether or not the point at issue is a controversial one, as it is in the case of the offshore workforce. Accordingly I am not concerned with the merits of the recognition of trade unions offshore or with the means by which support for such recognition should be ascertained. I have to concern myself with the question of safety, and in doing so take account of the existing situation in the North Sea.

21.84 In the light of the evidence which I have heard, which admittedly came almost entirely from trade union witnesses, I am prepared to accept that the appointment of offshore safety representatives by trade unions could be of some benefit in making the work of safety representatives and safety committees effective, mainly through the credibility and resistance to pressures which trade union backing would provide.

21.85   However, the position offshore is complicated by a number of factors: trade union membership is still relatively limited in relation to the total offshore workforce; trade unions have been "recognised" only to a limited extent; and the employment of offshore workers is fragmented between a number of different employers, with a high proportion being employed by contractors. As matters stand it does not seem to me to be appropriate to replace the 1989 Regulations with the offshore extension of the 1977 Regulations. This would remove safety representatives from a very large part of the workforce and would undo the limited progress which was achieved in difficult circumstances by the making of the 1989 Regulations. Further those regulations have been in force for only a short period. Experience will show whether or not representatives elected under those regulations lack adequate credibility or resistance to pressures. In the meantime I consider that it would be inappropriate for me to recommend any change in the method by which safety representatives are chosen. I understand that the regulatory body intends to review the 1989 Regulations after 2 years' experience of their working. When carrying out that review the regulatory body may consider that there is room for improving the effectiveness of safety representatives; and putting the trade unions' contentions to the test for that purpose. For example, it may consider that it is appropriate to modify the existing scheme so as to require that safety representatives are appointed by trade unions in certain cases, such as where a trade union had achieved recognition in relation to a substantial aspect of labour relations and had a substantial membership on the installation in question.

21.86   For the present I am satisfied that it is appropriate that the type of protection provided in the case of trade union activities under Sec 58(1)(b) of the Employment Protection (Consolidation) Act 1978 should also be afforded to the activities of an employee as a safety representative. The Trade Union Group also submitted that intimidation and the breaking of a contract should become a criminal offence where it was directed against the raising or pursuing of a complaint relating to health and safety. As regards any wider measures I consider that the correct course in the first instance is to look to the safety representative as the channel through whom complaints in regard to health and safety should be expressed. I am also aware of the efforts which the Secretary of State for Energy and UKOOA have made in order to demonstrate that victimisation is not to be tolerated and that the reporting of incidents affecting safety is to be encouraged.

21.87   The Trade Union Group and other parties made a number of specific criticisms of the 1989 Regulations. Since these regulations have only recently been introduced I do not in general think that it is appropriate for me to recommend alterations. However, there is one exception to that. Reg 27 provides that it is to be the duty of the employer of a safety representative to ensure that he is provided with such training in aspects of the function of a safety representative as may be reasonable in all the circumstances and that the employer is to meet any reasonable costs associated with such training including travel and subsistence costs. In the light of the evidence I consider that the burden of providing the training and bearing its cost should fall not on the employer but on the operator of the installation where the safety representative serves. The operator has a knowledge and a responsibility for safety on an installation which is far wider than that of contractors working on it. In the case of smaller contractors who may have few personnel working on an installation they may, as Mr Lyons suggested, have great difficulty in providing training for any of their employees who may be elected as a safety representative. It is extremely important that the safety committee should include an adequate representation of contractors' employees.

Chapter 22

The Regulatory Body

Introduction

22.1   In this chapter I will give my views as to the body which should be the regulatory body for the future offshore safety regime.

22.2   This involves considering a question which was studied by the Burgoyne Committee who reached the conclusion that the DEn was capable of discharging the responsibility of a single government agency for offshore safety, provided that it was suitably strengthened and sought advice from other bodies on matters of common concern (6.6). Since 1980 this matter has not been reviewed. There have been important developments in regulatory techniques in both onshore and offshore regimes. There has been direct experience of the capabilities and approach of the DEn and the HSE. The industry is on the threshold of what on any view are major changes which have important implications for the qualities required of the regulatory body. In the light of these considerations and the evidence which I heard in Part 1 of the Inquiry I considered that it was appropriate that this question should be considered in Part 2.

22.3   It is right that I should emphasise at this point that the proper context for the question is the future offshore safety regime. Much of the evidence and submissions were concerned with what was said to be past failures or successes on the part of the DEn and the HSE. However, these are relevant only in so far as they throw light on the appropriate choice for the future. Further that choice should take into consideration the implications of change at this stage in the history of the offshore safety regime.

The reasons for the conclusions of the Burgoyne Committee

22.4   It is clear that the committee attached significance to the differences between the offshore industry and the rest of industry in the United Kingdom, particularly in respect of the differences in environment and the remoteness of operation. There was need for special treatment which called for "flexibility of approach, speed of reaction and individual treatment of each case" in dealing with the problems of the offshore industry. They said that speed of response and flexibility of approach were more likely from an organisation with only one industry whose safety matters were its concern (4.16-18). General satisfaction had been expressed with the way in which the PED had approached its task. This was attributed to the selection of well qualified and experienced personnel. The DEn (and its predecessors) had grown up with the offshore industry and was in the best position to understand it and its problems (4.13-15).

22.5   On the other hand there had been criticism of the HSE's involvement in offshore safety, apparently due to its "lack of expertise" in certain areas such as deep diving, petroleum engineering and structural engineering in a marine environment. The assimilation of the offshore inspectorate into the HSE would take some time to be achieved (4.13-14). An organisation with responsibility for the majority of industrial safety would tend to show greater rigidity and a slower response (4.18).

22.6   On the footing that the DEn was the chosen agency "it is unthinkable that DEn would ignore advice on general trends and practices onshore in formulating offshore safety policy" (4.11).

22.7   As noted earlier two members of the committee, Mr Lyons and Mr Miller, dissented, essentially on the ground that a government department which was substantially responsible for the direction and control of an industry should not in any way be responsible for the standards and enforcement of occupational health and safety in that industry.

The alternatives

22.8   I am in no doubt that as matters stand the choice lies between the DEn on the one hand and the HSE on the other, on the basis that in either case the body is suitably strengthened for the task ahead. I heard detailed closing submissions in regard to that choice. The Trade Union Group and the Piper Disaster Group submitted that the HSE should replace the DEn as the regulatory body. UKOOA made submissions as to the qualities which should be possessed by the regulatory body. It "should be a single authority which has appropriate competence and expertise". However UKOOA were neutral as to which body I should recommend. It may be noted that at the time of the Burgoyne Committee UKOOA supported the DEn as the regulatory body. The Contractors' Interests favoured the retention of the *status quo*. The DEn did not itself enter into this controversy but their counsel assisted me greatly by acting as *amicus curiae* at my request and set out full arguments against the proposal for replacement of the DEn by the HSE.

22.9   In what follows I will set out what I have derived from the evidence as to the nature and capabilities of each body; and as to the way in which each has approached the development and enforcement of regulatory control.

The Department of Energy

22.10   The DEn is in the position, which the Burgoyne Committeee considered to be of some significance, of being able to concentrate on the offshore industry which, has many special features. As one would expect the department has acquired a great deal of knowledge of the industry. It is regularly in contact with bodies which represent operators, contractors and the workforce. These bodies are consulted in regard to proposed legislation and participate in the discussion of future guidance and research.

22.11   On the other hand the comparatively small size of the Safety Directorate means that the prospects for promotion of its personnel are limited. This may well be a factor which has tended to affect recruitment and retention of personnel. It is clear from the evidence to which I have referred in Chapters 15 and 21 that in a number of areas the work of the Safety Directorate has been hampered by persistent under-manning. The problem does not seem to be due, at least in recent times, to a shortage of financial resources but to a difficulty in recruiting. Although I have noted the initiatives which are being taken, it seems unlikely that this chronic problem will be readily solved.

22.12   The comparatively small size of the Safety Directorate appears also to have been a factor restricting the scope of the in-house expertise which it could employ, with the result that it placed more reliance on the work of consultants and other bodies such as the HSE than it would have done if it were part of a larger body with greater shared resources. At the same time I should say that it was brought out clearly in evidence that the HSE is always ready to provide assistance to the Safety Directorate. The limitations on the Safety Directorate's own expertise have a practical significance, and particularly for the future. Three points may be mentioned. Firstly, I accept Mr Ferrow's comment that the inspectorate "do not seem to have such direct and straightforward access to all the areas of expertise that they might want". Secondly, these limitations are likely to affect the ability of the regulatory body to give prompt and authoritative responses. The Directorate appears to be short of in-house expertise in fire and explosion protection. I noted that Mr Brandie suggested that the apparent reluctance of the DEn to support a scenario-based approach to the design of fire protection stemmed from a shortage of expertise to assess such design. If goal-setting regulations are to be brought into existence there would require to be an entirely different level of expertise from the present. Thirdly, the present intention of the Safety Directorate is to rely on certifying authorities for the assessment of the hardware aspects of FSA; and as regards the assessment of management the directorate is clearly short of the required expertise (see para 21.22).

22.13 It was strongly represented by trade union witnesses, in line with the dissent from the report of the Burgoyne Committee, that the Safety Directorate lacked, or at any rate was perceived to lack, independence. Put another way, it was suggested that there was a conflict of interest between the objectives of the Safety Directorate on the one hand and the objectives of other parts of the DEn on the other. However, it was pointed out in response that in Norway a single body, the NPD, is in control of both exploitation of resources and of safety; although it was responsible to different Ministries in regard to those functions. It was also pointed out that in the case of the United Kingdom the PED had two reporting lines. One was to the Secretary of State for Energy, who was in turn responsible to Parliament. There is a clear Ministerial commitment to safety and the Safety Directorate exercise direct access to Ministers as occasion arises. The other was to the HSC, in accordance with the arrangements set up by the Government in the light of the views of the members of the Burgoyne Committee.

The DEn's approach to the development of regulatory control

22.14 I have already discussed in Chapter 21 the DEn's lack of progress on goal-setting regulations, the unduly restrictive nature of existing regulations and guidance notes and the restricted use of the HSWA for regulations (summarised at para 21.51). This does not show the "speed of response and flexibility of approach" which the Burgoyne Committee considered that the DEn were more likely to exhibit.

22.15 At para 17.22 I commented that prior to the disaster the DEn does not appear to have addressed the major hazards presented by hydrocarbon inventories. This is further illustrated by the history of its attitude to the CIMAH Regulations and the introduction of FSA. In October 1983 the HSE asked the DEn whether the requirements of the proposed CIMAH Regulations for the provision of a Safety Case and emergency plans would be appropriate for offshore situations. In reply Mr Petrie, as Head of Operations and Safety, in a letter dated 24 November 1983 stated that it was considered that existing legislation under the MWA already covered the proposed requirements. He went on: "Furthermore this department has policy initiation responsibility for all offshore oil and gas safety matters and advises the HSC on such policy matters. It is our intention to advise the HSC against any extension of onshore major hazard legislation to offshore installations, where the legal and practical provisions are considered satisfactory and are already far in advance of these contemplated by HSE." This was at a time when the Burgoyne Committee had already recommended that the DEn should encourage a systematic approach to safety assessments of structures and plant during design and construction, with the purpose of establishing agreed procedures (6.27); and when in the NCS risk evaluation on a quantitative basis was already required.

22.16 Questioned as to what was "already far in advance" Mr Petrie said that these words had been justified by the existence since 1975 of the Construction and Survey Regulations. He said that although certifying authorities did not deal with safety assessment as such they dealt with design and construction in accordance with codes. He also referred to a number of miscellaneous requirements of other offshore regulations, and said: "We know what the hazard is only too well with significant amounts of hydrocarbons offshore, so that, together with many other regulations that apply offshore, we broadly felt that the objective of the regulation was already in place, including the general requirement, under the general aegis of the HSWA, for employers to ensure that they have a safe place of work." However, in my view, the problem of major hazards is not one which can be dealt with simply by following codes; and the miscellaneous requirements were not components of a system which was intended to be able to handle such hazards. Mr Petrie did not appear to have realised from the CIMAH Regulations that the preparation of a Safety Case was valuable in imposing a discipline on manufacturers to show that they had identified the major hazards and created appropriate controls, although this was part of the background to the regulations and is explicitly set out in the guidance notes which relate to them. He did not appear to be certain whether it had been realised that management systems were an important element in the CIMAH Safety Case. He admitted that he had not kept up with the development of the Safety Case under the CIMAH Regulations.

22.17   The proposition that the offshore provisions were "far in advance" is at odds with the DEn's discussion document on FSA, work on which began in 1987. This stated, *inter alia*: "For some time the Safety Directorate has been concerned that reliance on good engineering practice, the application of approved standards and the certification and inspection regimes do not of themselves comprehensively identify and highlight the hazards and sequences of events that can lead to a major accident."; and referred to FSA as embracing "the whole spectrum of safety analysis techniques that can be brought together in a structured framework to make a major step forward in enhancing the overall safety of offshore installations." As I have already observed at para 17.26, the document makes no reference to the CIMAH Regulations.

22.18   The evidence demonstrates, in my view, a serious failure on the part of the DEn to address the regulatory requirements for dealing with the major hazards, whether they arose from collisions or from a failure in pressure systems or in some other way. The result, as I said in para 21.54, has been to set back the development of the offshore safety regime by many years. The DEn's attitude appears to have been based in part on a failure to realise that the existing offshore provisions were not enough; and in part on a failure to understand the CIMAH Regulations and the Safety Case - a failure which, at least in the case of Mr Petrie, persisted throughout his evidence.

22.19   In about 1985 the DEn advised against the offshore application of the COSHH Regulations on the basis that the provisions of the Operational Safety, Health and Welfare Regulations were adequate. The COSHH Regulations represented a major change onshore, described by Mr Rimington as the most important reform for 13 years. Their basic aim was to ensure that where employees might be exposed to toxic substances there should be formal procedures to ensure that their exposure was minimised and was in any case kept below the maximum exposure level. Mr Petrie said that probably the main offshore provision which was relevant was Reg 4 of the Operational Safety, Health and Welfare Regulations, but it is clear that there is no true similarity between them. There was no evidence that the COSHH Regulations were unsuitable for offshore application; and indeed Mr McKee gave evidence that Conoco (UK) Ltd had unilaterally applied them to their installations.

22.20   The approach of the DEn seemed to me to tend towards over-conservatism, insularity and a lack of ability to look at the regime and themselves in a critical way. From this certain practical results have followed; the introduction of improvements in safety has been hampered; and the development of legislation on the basis of the HSWA has been kept back.

22.21   It does not appear to be perceived by the DEn that a radical change of approach is already due. Nothing appears to have been learnt from the experience of the NPD with which the DEn were in regular contact. Despite arrangements which should have enabled the DEn to obtain a wider view of modern approaches to the regulation of industrial safety, such as their relationship with the HSC, their work on the OIAC and their opportunities for exchange of ideas and personnel with the HSE, the offshore approach to the management of safety seems to me to be a number of years behind the approach onshore.

The DEn's approach to enforcement of regulatory control

22.22   I have already commented in Chapters 15 and 21 on the type of inspection practised by the DEn; and the absence of any systematic approach to the scrutiny of systems for the management of safety. Their approach appeared to me to be at least in origin mainly reactive; moves towards a more pro-active approach appear to have been slow and tentative. Here again my remarks in para 22.21 apply.

The Health and Safety Executive

22.23   The HSE encompasses responsibility in regard to both general and specialised industry onshore. It represents the principal source of safety expertise in the United Kingdom. While it employs consultants in many areas it clearly regards in-house expertise as essential. This expertise includes the fields of major hazards, safety assessment and the assessment of management systems. Mr Ferrow described the HSE as being "a fairly comprehensive and diverse organisation that allows individual inspectors in the field relatively quickly to get very expert advice on almost any matter you can think of".

22.24   The HSE has, of course, no current expertise or experience in regard to offshore installations. However, while the environment and remoteness of offshore installations are unlike anything onshore, the nature of the operations carried on offshore are no more complex than what may be encountered onshore. Further, as I have already indicated in Chapter 17 I see no reason why the onshore approach to major hazards and safety assessment should not, *mutatis mutandis*, be capable of extension to the offshore. In passing I should note that Mr Rimington, when asked to comment on the HSE's "lack of expertise" (see para 22.5), pointed out that the HSE had had no involvement with offshore engineering and would not have questioned the expertise of the DEn in regard to deep diving.

22.25   The HSE has not been without difficulties in achieving adequate recruitment. In regard to specialist inspectors Mr Rimington said that following the Chernobyl disaster there had to be a substantial pay rise in order to recruit additional nuclear inspectors. There was a current difficulty in retaining specialist inspectors in the Technology Division but the HSE was acting on advice which had been obtained in order to deal with this. A result of the HSE being the principal source of expertise was the 'poaching' of experts by industry. He was heartened by the fact that for 3 successive years the Government had given the HSC their full bid for financial provision.

22.26   It is clear that the HSE has always encouraged upward mobility among its personnel. It has recently elaborated a strategy for the development of careers in its organisation. This has also benefited recruitment and the retention of personnel. Mr Rimington said that inspectors were transferred into the specialist inspectorate from other parts of the HSE if they were suitably qualified. "So certainly we have transferred people, and more particularly ideas, from one part of the executive to another."

22.27   Mr Rimington said that the HSE's 'clout' with industry was in part due to its independence and in part to the fact that both sides of industry were represented on the HSC and in the working groups with which the HSE was closely involved.

The HSE's approach to the development of regulatory control

22.28   As I have stated in para 16.36 the HSE has made substantial progress with the modernisation of existing onshore legislation relating to health and safety. Progress has necessarily been slow owing to the need to formulate a new style of regulations for industry at large or for cases in which a special regime was required; and owing to the need to carry out consultation with a view to arriving at a consensus. Latterly the speed of development of legislation has been increased. HSE has plainly built up a strong body of knowledge and experience in the formulation of legislation which fulfils the policy of the HSWA and make major advances in techniques for the regulation of safety.

22.29   While one of the ways in which the HSE seeks to enforce a regulatory control is by inspection there appear to me to be significant differences in its approach to inspection compared with that of the DEn. It will be recalled that in para 16.37 I referred to the evidence of Mr Rimington that "an inspector's immediate purpose in visiting is to satisfy her or himself that systems exist that are likely to lead to the identification and prevention by management of significant faults and that the attitude of management is conducive to this." At para 21.24 I referred to the HSE's insistence on evidence of the monitoring of safety. This demonstrated a greater attention to the systems by which accidents can be prevented and mitigated. This effort is supported by the work of the APAU (see paras 16.39 and 21.61).

22.30   It was also clear from the evidence that the HSE have given a higher profile to the subject of safety both with the public and with industry. Plans and details of performance are published. Mr Rimington said: "If safety is cost effective, then in my view a high profile for it is cost effective."

Implications of change

22.31   In regard to the HSE, in view of its lack of existing expertise and experience in the offshore industry Mr Rimington was circumspect when commenting on the proposition that the HSE should become the regulatory body offshore - a change which had not been sought either by that body or the HSC. He emphasised the distinctive culture of the offshore industry and the importance that the regulated had confidence in the regulator. Such a change would call for flexibility and understanding on both sides. He pointed out the difficulties involved in organisational distribution and the care which would require to be taken to make sure that changes in the legislation occurred at no greater speed than they could be adapted to. On the other hand I am satisfied that there is no incompatibility between the offshore safety regime and the principles on which the onshore safety regime is presently organised. This includes the certification system which has no exact counterpart onshore. Further I have no doubt that the HSE has the necessary basic expertise for assuming responsibility offshore, although it is obvious that the HSE could not be expected to proceed without the assistance of PED inspectors and their accumulated knowledge and experience of the offshore industry. I should also point out that, as between these two bodies, I consider that the HSE would have the capacity to cope with the major management workload involved in the assumption of responsibility for offshore safety along with the other major changes which I have recommended for the future regime.

22.32   The transfer of responsibility to the HSE would bring to an end the agency agreement the operation of which has, in my view, a number of features which are not entirely satisfactory. Under that agreement the HSC as principal has no say in the quality or efficiency of the work of the DEn in regard to matters which fall within that agreement. In para 21.49 I pointed out Mr Rimington's comments on the difficulty of determining how far the HSC or the HSE were entitled to form judgements of PED's effectiveness. Finally this should assist in bringing to an end the tension between the MWA and the HSWA in the offshore safety regime.

22.33   In regard to the PED Mr Priddle pointed out that the transfer of responsibility from the PED to the HSE seemed likely to include some constraint on the free flow of information and some duplication of resources as between the Safety Directorate and other petroleum specialists in the PED. He thought that there would be some loss of career development opportunities and management flexibility.

Conclusion

22.34   I have considered carefully the factors which I have attempted to set out in the preceding paragraphs. I have come to the conclusion that the balance of advantage in the interests of the future offshore safety regime lies in favour of the transfer of responsibilities from the PED to the HSE. The decisive considerations in my mind arise from considering the differences in approach between these two bodies to the development and enforcement of regulatory control. These differences have been plain for some years and flow from differences in the way in which the bodies are directed and managed. I am confident that the major changes which I have recommended are ones which are in line with the philosophy which the HSE has followed. This alternative is clearly preferable to the PED even if it was given a higher level of manning with greater in-house expertise. I also attach importance to the benefits of integrating the work of the offshore safety regulator with the specialist functions of the HSE.

22.35 I am conscious that the change which I have recommended will take some time to implement and will inevitably involve disruption. Successful implementation will call for co-operation, flexibility and understanding at all levels between the industry and the existing and future regulatory bodies. Special treatment will be required in regard to certain functions which are presently discharged by the Safety Directorate. These include the planning, as distinct from the safety, functions in regard to offshore pipelines; and the function of administering well consents. It is appropriate that these functions should be retained by the DEn.

22.36 As regards the flow of information between specialists in the PED, I noted earlier that the Safety Directorate is consulted about and can express reservations on safety grounds in regard to important stages in the licensing process (see para 16.18). It is clear that in order to do this effectively they must be involved in and aware of the discussions between specialists in the EADU and operators in regard to both exploration and development. It is of major importance, in my view, that such links between the EADU and the regulatory body for safety be maintained in the transfer of responsibility for safety to the HSE.

22.37 While offshore safety stands to benefit by responsibility being transferred to the HSE it is important that the distinctive character and requirements of the offshore industry should be recognised in the administrative arrangements within the HSE. For this reason it is also my view that responsibility for offshore safety should be discharged by a discrete division of the HSE which is exclusively devoted to offshore safety and is able to respond promptly and authoritatively to its special needs. This division should employ a specialist inspectorate and should have a clear identity and strong influence in the HSE. It should be headed by a chief executive who should be responsible directly to the Director General of the HSE and should be a member of its senior management board. His function would include the development of the offshore safety regime, and in particular the implementation of its provisions for Safety Cases and SMS. The need for adequate resources in order to meet these changes is obvious.

22.38 In these circumstances there is little which I require to say in regard to the complaint that the Safety Directorate is not independent or perceived to be independent and accordingly is not well fitted to carry out the functions of the regulatory body in regard to safety matters. On the evidence I was not convinced that the Safety Directorate actually lacks independence or that its actions had been affected by considerations related to the exploitation of resources. On the other hand there is a perception, at least among some trade unionists, that it lacks independence. This is an unfortunate feature of the present scene. However, if my recommendations in this chapter are followed it will no longer be a live issue.

Chapter 23

Recommendations

In this chapter I will set out my recommendations in the light of the matters discussed in Chapters 17-22. Each recommendation is followed by reference to the paragraph in the earlier chapter to which it is directly related. The recommendations are arranged according to the following subjects:-

Safety Case

1. The operator should be required by regulation to submit to the regulatory body a Safety Case in respect of each of its installations. The regulation should be analogous to Reg 7 of the CIMAH Regulations, subject to recommendations 2–13 (paras 17.33-43).

2. The Safety Case should demonstrate that certain objectives have been met, including the following:-

   (i)   that the safety management system of the company (SMS) and that of the installation are adequate to ensure that (a) the design and (b) the operation of the installation and its equipment are safe (paras 17.36 and 21.56-57);

   (ii) that the potential major hazards of the installation and the risks to personnel thereon have been identified and appropriate controls provided (para 17.37); and

(iii) that adequate provision is made for ensuring, in the event of a major emergency affecting the installation (a) a Temporary Safe Refuge (TSR) for personnel on the installation; and (b) their safe and full evacuation, escape and rescue (paras 17.37-38, 19.109, 19.157 and 20.8).

3. The SMS should be in respect of (a) the design (both conceptual and detailed) of the operator's installations; and (b) the procedures (both operational and emergency) of those installations. In the case of existing installations the SMS in respect of design should be directed to its review and upgrading so far as that is reasonably practicable (para 21.56).

The SMS should set out the safety objectives, the system by which these objectives are to be achieved, the performance standards which are to be met and the means by which adherence to these standards is to be monitored (para 21.56).

It should draw on quality assurance principles similar to those stated in BS 5750 and ISO 9000 (para 21.58).

4. In furtherance of the objectives set out in para 2 above, the operator should be required to set out the following in the Safety Case:-

(i) A demonstration that so far as is reasonably practicable hazards arising from the inventory of hydrocarbons

(a) on the installation, and

(b) in risers and pipelines connected to the installation both in themselves and as components of the total system of which they form part

have been minimised (paras 19.17 and 19.20).

(ii) A demonstration that so far as is reasonably practicable the exposure of personnel on the platform to accidental events and their consequences has been minimised (para 17.37).

(iii) A demonstration by quantified risk assessment of major hazards that the acceptance standards have been met in respect of risk to the integrity of the TSR, escape routes, embarkation points and lifeboats from design accidental events and that all reasonably practicable steps have been taken to ensure the safety of persons in the TSR and using escape routes and embarkation points (paras 17.38 and 19.157).

(iv) A demonstration that within the TSR there are facilities as specified by the operator which are adequate for the purpose of control of an emergency (para 19.182).

(v) A fire risk analysis, in accordance with recommendation 49 below (para 19.90).

(vi) An evacuation, escape and rescue analysis, in accordance with recommendations 73-75 below (para 20.9).

5. For the purposes of the demonstration referred to in para (iii) of recommendation 4, the accidental events are to be identified by the operator. A design accidental event is an event which will not cause the loss of any of the following:-

— the integrity of the TSR,

— the passability of at least one escape route from each location on the platform,

— the integrity of a minimum complement of embarkation points and lifeboats specified for personnel in the TSR, and

— the passability of at least one escape route to each of these embarkation points,

within the endurance period specified. Events more severe than this are referred to as residual accidental events (para 19.160).

The acceptance standards for risk and endurance time should be set before the submission of the Safety Case. Standards should be set by reference to the ALARP principle. For the time being it should be the regulatory body which sets these standards. The operator should define the conditions which constitute loss of integrity of, and the standards of protection for, the TSR and escape routes to the TSR and from the TSR to the embarkation points; and should specify the minimum complement of embarkation points and lifeboats for the TSR (paras 19.158-159).

6. The TSR should normally be the accommodation (paras 19.156 and 19.161).

In the case of existing installations any requirement for the upgrading of the accommodation, escape routes and embarkation points should be determined on the basis of the Safety Case (para 19.165).

7. In connection with the above the Safety Case should specify the following:-

In respect of the TSR-

— its function
— the conditions which constitute its integrity
— the conditions for integrity of its supporting structure
— the events in which and the period for which it is to maintain its integrity (paras 19.157-158).

In respect of escape routes to the TSR and from the TSR to the embarkation points-

— the conditions which constitute their passability
— the conditions for integrity of their supporting structure
— the events in which and the periods for which they are to maintain their passability (*provided* that for each location on the platform there should be a minimum of two escape routes to the TSR, at least one of which should remain passable for the period) (para 19.164).

In respect of embarkation points and lifeboats-

— the number and location
— the conditions for their integrity and that of their supporting structure
— the events in which and the periods for which they are to maintain their integrity
— the minimum complement for the TSR (para 19.164).

8. No fixed installation should be established or maintained in controlled waters; and no mobile installation should be brought into those waters with a view to its being stationed there or maintained in those waters unless a Safety Case in respect of that installation has been submitted to and accepted by the regulatory body (para 17.41).

9. As regards existing installations the date for submission of the Safety Case should be laid down by regulation. There is an urgent need for the submission of Safety Cases, but the date should be selected by the regulatory body. The regulatory body should have the power, in the event of the failure of an operator to submit an acceptable Safety Case, to require the operator to take whatever remedial action it considered necessary, including requiring the installation to be shut down (paras 17.44-45).

10. A Safety Case should be updated:-

(i) After a period of years from its last assessment (not less than 3, not more than 5, years).

(ii) At the discretion of the regulatory body on the ground of a material change of circumstances, such as a change of operator, the occurrence of a major emergency (including one in which there is a precautionary evacuation), a major technological innovation or the discovery or better understanding of a major hazard.

However, provision should be made in order to avoid the need for more than one Safety Case to be updated by an operator at the same time; and to enable the regulatory body to postpone the automatic updating where it has recently required a discretionary updating (para 17.46).

11.   As regards modifications to installations or their equipment or procedures, the operator should, before putting the modification into effect, ascertain what effect, if any, it has on the relevant components of the Safety Case. An operator should be required to report to the regulatory body all intended modifications which meet criteria set by the regulatory body, with a view to discussing with the regulatory body whether and to what extent a review of the Safety Case is required (para 17.47).

12.   For the time being the acceptance by the regulatory body of Safety Cases should not be regarded as justifying the revocation of regulations or the withdrawal of guidance notes (para 17.67).

Where an operator proposes to meet the objectives of a Safety Case by means which are not in accordance with regulations or guidance notes the justification for such a course should be set out in the Safety Case. For the assistance of operators the regulatory body should publish as soon as possible, and thereafter update in the light of experience, a list of the individual regulations relating to an installation and its equipment in respect of which it is prepared to grant exemption in the light of a satisfactory demonstration in a Safety Case; and to do likewise in regard to guidance notes (para 17.67).

In due course the existing regulations of a detailed prescriptive nature should be reviewed with a view to their revocation or replacement by regulations which set objectives. However, it is anticipated that there will continue to be even in the long term a case for some detailed prescriptive regulations (paras 17.63, 17.67 and 21.67).

13.   The regulatory body should discuss with the industry whether it is desirable and practicable that at the stage of the application for Annex B consent (or its equivalent) there should be a procedure for submission by operators of a preliminary assessment of matters relevant to a Safety Case and for the acceptance of this assessment being a prerequisite for the granting of Annex B consent (para 17.43).

Auditing of the operator's management of safety

14.   The operator should be required to satisfy itself by means of regular audits that its SMS is being adhered to (para 21.60).

15.   The regulatory body should be required regularly to review the operator's audit on a selective basis; and itself to carry out such further audit as it thinks fit; and by regular inspection verify that the output of the SMS is satisfactory (para 21.60).

Independent assessment and surveys of installations

16.   The regulatory body should consider (i) after the introduction of requirements for the demonstration of SMS and auditing of compliance with it; and (ii) after experience in the operation and effectiveness of such requirements whether and to what extent it will be appropriate to retain the present system of certification (para 21.64).

Legislation - General

17.   (i) The principal regulations in regard to offshore safety should take the form of requiring that stated objectives are to be met (referred to as "goal-setting

regulations") rather than prescribing that detailed measures are to be taken (para 21.67).

(ii) In relation to goal-setting regulations, guidance notes should give non-mandatory advice on one or more methods of achieving such objectives without prescribing any particular method as a minimum or as the measure to be taken in default of an acceptable alternative (para 21.67).

(iii) However, there will be a continuing need for some regulations which prescribe detailed measures (para 21.67).

18. The provisions of the Mineral Workings (Offshore Installations) Act 1971 and the Petroleum and Submarine Pipe-lines Act 1975 which have the same general purposes as those of Part 1 of the Health and Safety at Work etc Act 1974 (HSWA), and the regulations made under such provisions, should be made relevant statutory provisions for the purposes of the HSWA (para 21.68).

19. The Construction and Survey Regulations, the Fire Fighting Equipment Regulations, the Life-Saving Appliances Regulations and the Emergency Procedures Regulations should be revoked and replaced by-

(i) Construction Regulations, covering *inter alia* the structure and layout of the installation and its accommodation.

(ii) Plant and Equipment Regulations, covering *inter alia* plant and equipment on the installation and in particular those handling hydrocarbons.

(iii) Fire and Explosion Protection Regulations, covering *inter alia* both active and passive fire protection and explosion protection, and

(iv) Evacuation, Escape and Rescue Regulations, covering *inter alia* emergency procedures, life-saving appliances, evacuation, escape and rescue.

Each of the above sets of regulations should include goal-setting regulations as their main or primary provisions and should be supported by guidance notes giving advice which is non-mandatory in the sense set out in paragraph (ii) of recommendation 17 (para 21.69).

20. Operators should be encouraged to specify the standards which they will use to comply with goal-setting regulations. For a given installation compliance may be demonstrated by reference to such standards, the terms of guidance notes and what is shown by a safety assessment or a combination of one or more of such methods (paras 17.66 and 21.70).

21. As regards existing guidance notes the regulatory body should consider whether and to what extent they should be treated without replacement or modification as giving non-mandatory advice in the sense set out in paragraph (ii) of recommendation 17; and should inform the industry accordingly (para 21.71).

22. In connection with the preparation of guidance notes the regulatory body should review the procedures for consultation so as to ensure that the views of the representatives of employers and employees involved in work offshore are adequately taken into account (para 21.72).

The regulatory body

23. There should be a single regulatory body for offshore safety (para 21.62).

24. The single regulatory body should discharge the safety functions in relation to fire-fighting equipment and life-saving appliances. As regards standby vessels it should discharge all functions, whether directly or through the agency of the Department of Transport (DoT), save those which relate to the statutory responsibility of the DoT under the Merchant Shipping Acts (paras 21.65-66).

25. The functions of the Petroleum Engineering Division of the Department of Energy (DEn) which are concerned with the regulation of offshore safety should in future be discharged by a discrete division of the Health and Safety Executive (HSE) which is exclusively devoted to offshore safety (paras 22.34 and 22.37).

26. This division should employ a specialist inspectorate and have a clear identity and strong influence in the HSE. It should be headed by a chief executive who should be responsible directly to the Director General of the HSE and should be a member of its senior management board. His function would include the development of the offshore safety regime, and in particular the implementation of its provisions for Safety Cases and SMS (para 22.37).

Safety committees and safety representatives

27. The regulatory body, operators and contractors should support and encourage the involvement of the offshore workforce in safety. In particular, first line supervisors should involve their workforce teams in everyday safety (para 18.48).

28. The operator's procedures included in line management of operations which are aimed at involving the workforce in safety should form part of its SMS (para 21.56).

29. The DEn's intention to review the Offshore Installations (Safety Representatives and Safety Committees) Regulations 1989 after 2 years' experience of their working is endorsed (para 21.85).

30. Safety representatives should be protected against victimisation by a provision similar to Sec 58(i)(b) of the Employment Protection (Consolidation) Act 1978 (para 21.86).

31. The Offshore Installations (Safety Representatives and Safety Committees) Regulations 1989 should be modified to the effect that the training of safety representatives should be determined and paid for by the operator (para 21.87).

Permits to work

32. The operator's permit to work system should form part of its SMS (para 21.56).

33. Operators and the regulatory body should pay particular attention to the training and competence of contractors' supervisors who are required to operate the permit to work system (paras 18.17 and 18.29).

34. Standardisation of the permit to work system throughout the industry is neither necessary nor practicable. However, in view of the fact that there is much in common between the systems of different operators, the industry should seek to increase harmonisation, for example in the colours used for different types of permits to work and in the rules as to the period for which a permit to work remains valid (para 18.28).

35. While it is not inappropriate for contractors' supervisors to act as Performing Authorities, operators should be made responsible for ensuring that such supervisors are trained in the permit to work system for the installation where they are to act as Performing Authorities and that they carry documentary proof of having completed such training (para 18.29).

36. All permit to work systems should incorporate a mechanical isolation procedure which involves the physical locking off and tagging of isolation valves (para 18.29).

37. A permit to work and its consequent isolations, both mechanical and electrical, should remain in force until the work is sufficiently complete for the permit to be signed off and the equipment returned to operation (para 18.8).

38. Copies of all issued permits to work should be displayed at a convenient location and in a systematic arrangement such that process operating staff can readily see and check which equipment is under maintenance and not available for operation (para 18.8).

Incident reporting

39. The regulatory body should be responsible for maintaining a database with regard to hydrocarbon leaks, spills and ignitions in the industry and for the benefit of the industry. The regulatory body should:-

    (i)   discuss and agree with the industry the method of collection and use of the data,

    (ii)   regularly assess the data to determine the existence of any trends and report them to the industry, and

    (iii) provide operators with a means of obtaining access to the data, particularly for the purpose of carrying out quantified risk assessment (para 18.43).

Control of the process

40. Key process variables, as determined by the Safety Case, should be monitored and controllable from the Control Room (para 18.36).

41. The Control Room should at all times be in the charge of a person trained and qualified to undertake the work of Control Room operator. The Control Room should be manned at all times (para 18.35).

42. The training of Control Room operators should include instruction in an onshore course in the handling of emergencies (para 18.35).

Hydrocarbon inventory, risers and pipelines

43. The Emergency Pipe-line Valve Regulations should continue in force until they are subsumed in the Plant and Equipment Regulations. The provision in these regulations for there to be on each riser a valve with full emergency shutdown capability and located as close to sea level as practicable is endorsed (paras 19.34-35).

44. There should be no immediate requirement that a subsea isolation valve (SSIV) be fitted on a pipeline connected to an installation. The operator should demonstrate in the Safety Case that adequate provision has been made, including if necessary the use of SSIVs, against hazards from risers and pipelines (para 19.36).

45. Studies should be carried out with the following objectives:-

    (i)   To explore the feasibility of dumping in an emergency large oil inventories, such as those in the separators, in a safe and environmentally acceptable manner, so as to minimise the inventory of fuel available to feed a fire (para 19.19).

    (ii)   To minimise the pipeline connections to platforms (para 19.21).

46. Studies should be carried out with the following objectives:-

    (i)   To achieve effective passive fire protection of risers without aggravating corrosion (para 19.22).

    (ii)   To improve the reliability and reduce the cost of SSIVs so that it is more often reasonably practicable to install them (para 19.37).

Fire and gas detection and emergency shutdown

47. The arrangements for the activation of the emergency shutdown valves (ESVs), and of SSIVs if fitted, on pipelines should be a feature of the Safety Case (para 19.42).

393

48. Studies should be done to determine the vulnerability of ESVs to severe accident conditions and to enhance their ability to survive such conditions (para 19.43).

Fire and explosion protection

49. Operators should be required by regulation to submit a fire risk analysis to the regulatory body for its acceptance (para 19.90).

50. The regulations and related guidance notes should promote an approach to fire and explosion protection:-

   (i)   which is integrated as between -

   — active and passive fire protection

   — different forms of passive fire protection, such as fire insulation and platform layout, and

   — fire protection and explosion protection (paras 19.87-95);

   (ii)   in which the need for, and the location and resistance of, fire and blast walls is determined by safety assessment rather than by regulations (para 19.96);

   (iii) in which the function, configuration, capacity, availability and protection of the fire water deluge system is determined by safety assessment rather than by regulations (paras 19.97 and 19.99);

   (iv)   which facilitates the use of a scenario-based design method for fire protection as an alternative to the reference area method (paras 19.91 and 19.98); and

   (v)   which provides to a high degree the ability of the fire water deluge system, including the fire pump system, to survive severe accident conditions (para 19.100).

51. The ability of the fire water deluge system, including the fire pump system, to survive severe accident conditions should be a feature of the Safety Case (para 19.100).

52. The regulatory body should work with the industry to obtain agreement on the interpretation for design purposes of its interim hydrocarbon fire test and other similar tests. If in the view of the regulatory body there exists a need for an improved test, such as a heat flux test, it should work with the industry in order to develop one (para 19.101).

53. The DEn discussion document on Fire and Explosion Protection should be withdrawn (para 19.102).

54. The regulatory body should ask operators which have not already done so to undertake forthwith a fire risk analysis, without waiting for legislation (para 19.103).

Accommodation, TSR, escape routes and embarkation points

55. Provisions should continue to be made by regulations supported by guidance notes as to the construction of the accommodation; and as to escape routes and embarkation points (para 19.166).

56. The regulations and the related guidance notes should promote an approach to protection of the accommodation:-

   (i)   in which external fire protection is provided both to prevent breach of the accommodation and to maintain breathable air within it (para 19.170); and

   (ii)   in which an integrated set of active and passive measures is provided to prevent ingress of smoke and other contaminants into the accommodation and to maintain breathable air within it (paras 19.170-171).

57. For the purpose of maintaining breathable air within the accommodation, it should be required by regulation that the ventilation air intakes should be provided with smoke and gas detectors and that on smoke or gas alarm the ventilation and dampers should shut down (para 19.172).

58. The regulations and related guidance notes on escape routes should recognise that it may not be practicable to protect escape routes against all physical conditions; and accordingly should be based on the objective that they should remain passable (para 19.174).

59. It should be required by regulation that escape routes are provided with adequate and reliable emergency lighting and with photoluminescent direction signs (para 19.175).

60. The regulatory body should ask operators which have not already done so to carry out forthwith an assessment of the risk of ingress of smoke or gas into the accommodation; and to fit smoke and gas detectors and implement ventilation shutdown arrangements as in recommendation 57, without waiting for legislation (para 19.173).

61. Studies should be carried out with the objective of assisting designers in predicting the breathability of air in a TSR where its external fire wall is subjected to a severe hydrocarbon fire (para 19.163).

Emergency centres and systems

62. It should be required by regulation that there should be available within the TSR certain minimum specified facilities for the monitoring and control of an emergency under hostile outside conditions (paras 19.178 and 19.182).

These facilities should be in the Control Room, which should be located in the TSR (para 19.179).

On existing installations where the Control Room is not in the TSR, these facilities should be in an Emergency Control Centre located in the TSR. In such a case the Control Room should be protected against fire and explosion as determined by safety assessment (paras 19.180-181).

63. It should be required by regulation that a Radio Room with facilities for external communications should be located in the TSR (para 19.179).

On existing installations where the Radio Room is not in the TSR, these facilities should be in an Emergency Radio Room located in the TSR (para 19.180).

64. The regulations and related guidance notes should promote an approach to emergency systems:-

   (i)   which provides to a high degree the ability of these systems to survive severe accident conditions (paras 19.188-189); and

   (ii) which applies to communications systems the fail-safe principle (para 19.193).

The emergency systems include the emergency power supplies and systems, the emergency shutdown system and the emergency communications systems. Severe accident conditions include fire, explosion and strong vibration (para 19.188).

65. The ability of emergency systems to survive severe accident conditions should be a feature of the Safety Case (para 19.189).

66. The regulatory body should work with the industry to promote the use of status light systems (para 19.192).

67. The regulatory body should work with the industry to achieve standardisation of status lights and of alarm systems for emergencies (para 19.194).

68.   Studies should be done to determine the vulnerability of emergency systems to severe accident conditions and to enhance their ability to survive such conditions (para 19.190).

69.   The regulatory body should ask operators which have not already done so to review forthwith the ability of emergency systems to withstand severe accident conditions (para 19.191).

70.   Where a regulation imposes a requirement for a major emergency or protective system, such as a fire deluge system, it should be required that the operator should set acceptance standards for its availability (para 19.199).

Pipeline emergency procedures

71.   Operators should be required by regulation regularly to review pipeline emergency procedures and manuals. The review should ensure that the information contained in manuals is correct, that the procedures contained are agreed with those who are responsible for executing them and are consistent with the procedures of installations connected by hydrocarbon pipelines (para 19.196).

72.   Operators should be required by regulation to institute and review regularly a procedure for shutting down production on an installation in the event of an emergency on another installation which is connected to the first by a hydrocarbon pipeline where the emergency is liable to be exacerbated by continuation of such production (para 19.197).

Evacuation, escape and rescue - General

73.   Operators should be required by regulation to submit to the regulatory body for its acceptance an evacuation, escape and rescue analysis in respect of each of its installations (para 20.9).

74.   The analysis should specify the facilities and other arrangements which would be available for the evacuation, escape and rescue of personnel in the event of an emergency which makes it necessary or advisable in the interests of safety for personnel to leave the installation (para 20.9).

75.   In particular the analysis should specify:-

(i)   The formal command structure for the control of an emergency affecting the installation;

(ii)   The likely availability and capacity of helicopters, whether in-field or otherwise, for the evacuation of personnel;

(iii) The types, numbers, locations and accessibility of totally enclosed motor propelled survival craft (TEMPSC) available for the evacuation of personnel from (a) the TSR and (b) other parts of the installation from which access to the TSR is not readily available;

(iv)   The types, numbers and locations of life rafts and other facilities provided as means of escape to the sea;

(v)   The specification (including speed, sea capability and accommodation), location and functions of the standby vessel and other vessels available for the rescue of personnel;

(vi)   The types, numbers, locations and availability of fast rescue craft, whether stationed on the installation or on the standby or other vessels; and

(vii) The types, numbers and locations of personal survival and escape equipment.

(All in para 20.9).

76. The regulatory body should ask operators which have not already done so to undertake an evacuation, escape and rescue analysis forthwith, without waiting for legislation. The timetable for completion of this analysis should be agreed between the regulatory body and the industry but should not exceed a total of 12 months, and that only for operators of a large number of installations (para 20.9).

Helicopters

77. Operators should adopt a flight following system for determining at short notice the availability and capacity of helicopters in the event of an emergency. This system could be either a system operated by the individual operator or a North Sea-wide system (para 20.11).

TEMPSC

78. The requirement by regulation that each installation should be provided with TEMPSC having in the aggregate sufficient capacity to accommodate safely on board 150% of the number of persons on the installation should be maintained (para 20.16).

Such provision should include TEMPSC which are readily accessible from the TSR and which have in the aggregate sufficient capacity to accommodate safely on board the number of persons on the installation (para 20.16).

79. On new installations where the provision of davit-launched TEMPSC is acceptable to the regulatory body they should be oriented so as to point away from the installation (para 20.24).

80. The regulatory body should work with the industry to develop equipment and methods to enable TEMPSC to be launched clear of the installation including where, as on existing installations, they are oriented so as to point along the side of the installation (para 20.18).

81. Reg 5 of the Life-Saving Appliances Regulations should be amended or replaced so as to enable free-fall TEMPSC to be installed on new and existing installations. It should remain for the operator to justify its choice of TEMPSC as being appropriate in the particular conditions of its installation (para 20.24).

Means of escape to the sea

82. It should be required by regulation that each installation should be provided with life rafts having in the aggregate sufficient capacity to accommodate safely on board at least the number of persons on board the installation; along with suitable ropes to enable those persons to obtain access to the life rafts after they have been launched and deployed (para 20.26).

83. A variety of means of descent to the sea should be provided on all installations. In accordance with recommendation 75 the types, numbers and locations of facilities for this purpose should be specified in the evacuation, escape and rescue analysis; but such facilities should include:-

— fixed ladders or stairways
— personal devices for controlled descent by rope (paras 20.28-29).

84 The regulatory body should work with the industry to determine the practicability and safety of escape chutes and collapsible stairways (para 20.30).

Personal survival and escape equipment

85. Each individual on board an installation should be provided with:-

(i) a personal survival (or immersion) suit;

(ii)   a life-jacket;

(iii) a smoke hood of a simple filter type to exclude smoke and provide protection for at least 10 minutes during escape to or from the TSR;

(iv)   a torch; and

(v)   fireproof gloves.

These articles should be kept in the accommodation (para 20.36).

Other survival suits, life-jackets and smoke hoods for at least one half of the number of persons on the installation should be stored in containers placed at suitable locations on the installation (para 20.36).

86.   The use of small transmitters or detectors on life-jackets in order to assist in the finding of personnel in the dark should be considered. Luminescent strips should be of a colour other than orange (paras 20.33-34).

87.   Work should be carried out with the objective of combining the functions of a survival suit and a life-jacket in one garment (para 20.32).

Standby vessels

88.   Changes in the regulations and the code for the assessment of standby vessels should be aimed at an improvement in the quality of standby vessels, introducing basic standards for existing vessels and higher specifications for new vessels (para 20.41).

89.   It should be required by regulations that each standby vessel should comply with the following standards:-

(i)   It should be highly manoeuvrable and able to maintain its position;

(ii)   It should provide full visibility of the water-line in all directions from the bridge;

(iii) It should have at least two 360° searchlights capable of being remotely controlled;

(iv)   It should have two fast rescue craft. One of the 2 fast rescue craft should be able to travel at 25 knots in normal sea states. The smaller fast rescue craft (9 person capacity) should be crewed by 2 persons; the larger by 3 persons. Fast rescue craft should be equipped with adequate means of communicating with the standby vessel by VHF radio; and carry an adequate portable searchlight;

(v)   It should have the means of rapid launching of its fast rescue craft;

(vi)   It should have adequate means of communication by radio with its fast rescue craft, the installation, nearby vessels and the shore; and

(vii) It should have at least two methods of retrieving survivors from the sea.

(All in para 20.42).

90.   Reg 10 of the Emergency Procedures Regulations should be revoked (para 20.39).

91.   Sec 3 of the code for the assessment of standby vessels (areas of operation) should be withdrawn (para 20.39).

92.   The owners of standby vessels should be required to notify the regulatory body weekly as to the locations and functions of their vessels in the ensuing week. A copy of such notification should also be given to the DoT (para 20.54).

93.   As regards the appropriate numbers for the crew of standby vessels, the DoT should take into account the evidence given in the Inquiry when reviewing the code in this respect (para 20.50).

94. The proposals in the amended code as to age limit, medical examination and certification of fitness of members of the crew of standby vessels; and as to their periods of duty are endorsed (paras 20.51-52).

95. The regulatory body should work with the industry to obtain agreement as to adequate training packages for the crew of standby vessels. Such training should be administered, and records of training kept by the Offshore Petroleum Industry Training Board (OPITB) (para 20.55).

96. The coxwain and crew of fast rescue craft should receive special training for their duties, along with regular refreshers (para 20.55).

Command in emergencies

97. The operator's formal command organisation which is to function in the event of an emergency should form part of its SMS (para 20.59).

98. The operator's criteria for selection of OIMs, and in particular their command ability, should form part of its SMS (para 20.59).

99. There should be a system of emergency exercises which provides OIMs with practice in decision-making in emergency situations, including decisions on evacuation. All OIMs and their deputies should participate regularly in such exercises (para 20.61).

Drills, exercises and precautionary musters and evacuations

100. The operator's system for emergency drills and exercises should form part of its SMS (paras 20.61 and 20.64).

101. Offshore emergency drills and exercises should be carried out in accordance with the UKOOA guidelines for offshore emergency drills and exercises on installations (paras 20.61 and 20.64).

102. All offshore staff should attend one muster per tour of duty (para 20.62).

103. The circumstances of all precautionary musters and evacuations should be reported by operators to the regulatory body (para 20.62).

104. Operators should maintain lists of personnel on board by alphabetical order and also by reference to the names of contractors whose personnel are represented on board. These lists should be updated for every movement of personnel and copied immediately to the shore (para 20.62).

Training for emergencies

105. The UKOOA guidelines for offshore emergency safety training on installations should be a minimum requirement for survival, fire-fighting and other forms of training detailed therein for the relevant personnel employed offshore. Personnel who have not met the requirements of these guidelines should not be permitted to work offshore (para 20.64).

In order to ensure that these guidelines are complied with operators should be required to devise and maintain a system for the purpose, pending the date when the central training register instituted by OPITB for recording the personal details and safety training courses attended by all personnel seeking employment offshore is fully operational (para 20.64).

106. The operator's system for emergency training and its enforcement should form part of its SMS (para 20.64).

Overleaf – Plates 1 & 2   Oil and Gas fields in the northern North Sea.

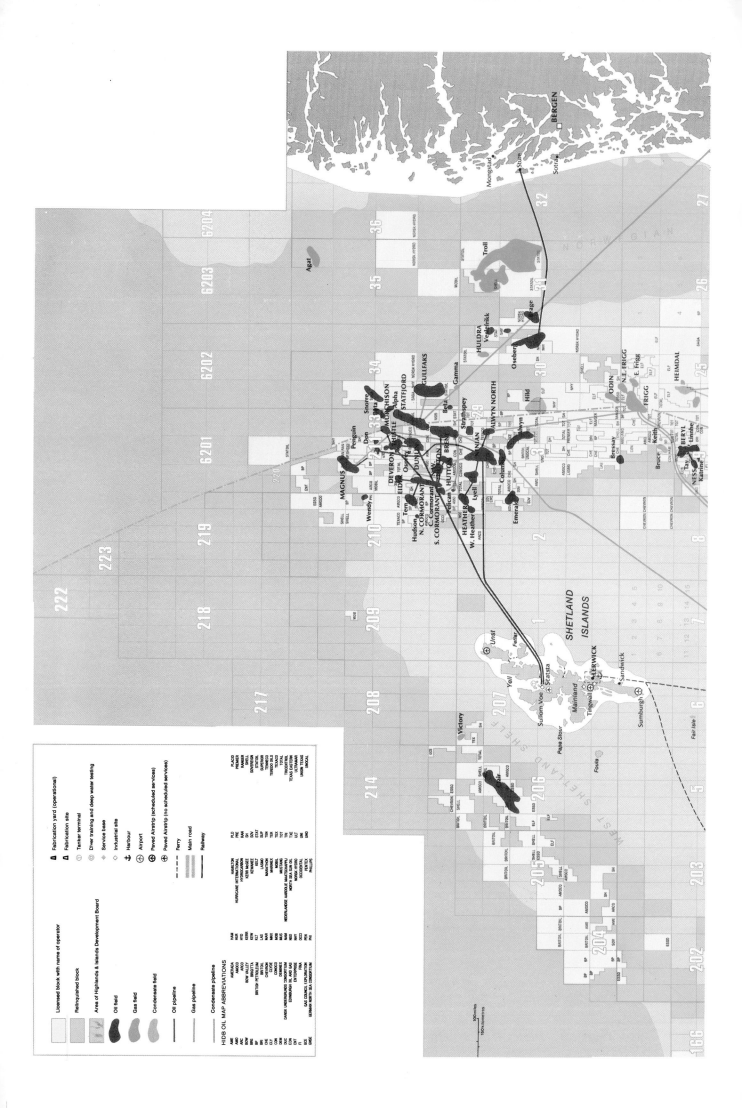

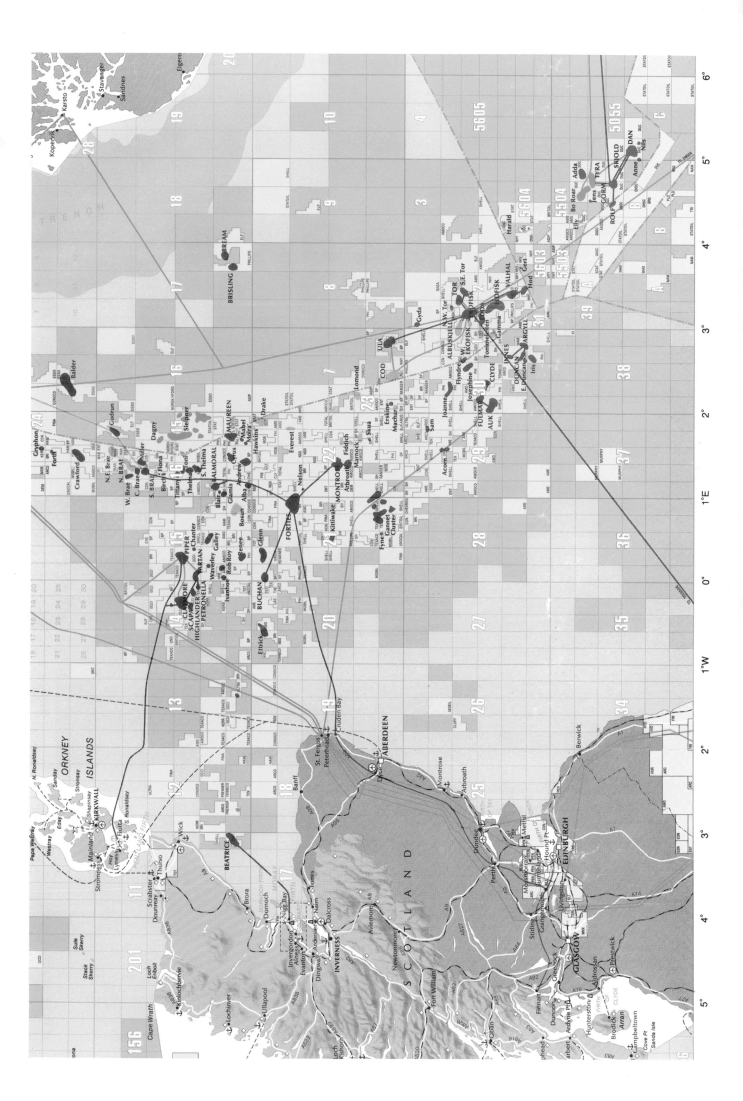

Plate 3   The Piper Alpha platform: view from the north-west.

Plate 4 The Piper Alpha platform: view from the south-east.

Plate 5   The Piper Alpha platform: west face and pipe deck.

Plate 6   Model of the production modules (1: 33 scale): B Module looking east.

Plate 7 Model of the production modules (1: 33 scale): C Module looking east.

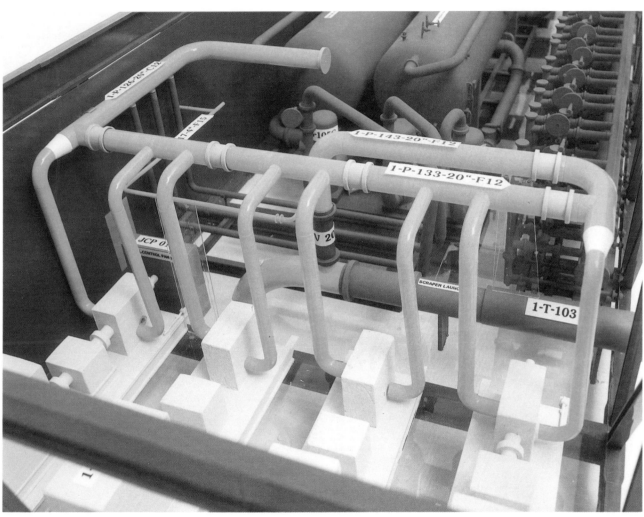

Plate 8 Model of the production modules (1: 33 scale): the main oil line pumps at the west end of B
Module. The 4 inch condensate injection line is the small yellow line 2–P–517–4″–F15 coming
through the B/C firewall and entering the MOL pump discharge header 1–P–143–20″–F12.

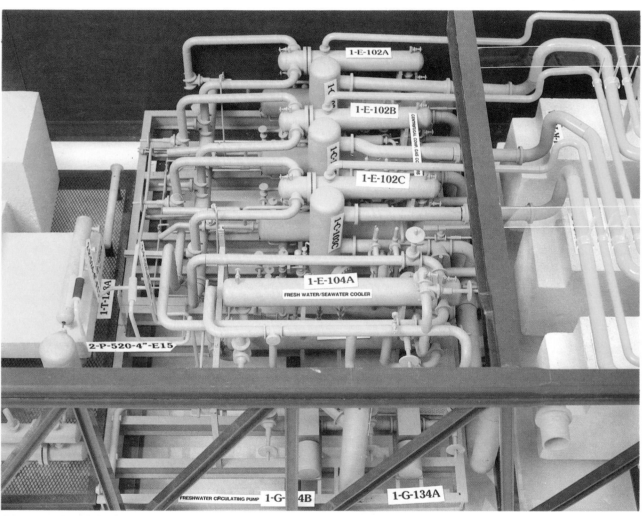

Plate 9 Model of the production modules (1: 33 scale): the centrifugal compressor skid towards the east end of C Module. The location of PSV 504 is marked in red.

Plate 10(a)   Control panels and control desk in the Control Room. The view is of the space between the control desk and the main process control panel, looking south (see Fig.J.4(c)). The operator is sitting with his left elbow on the control desk. To the east and going from south to north are the main fire and gas control panel and then the main process control panel.

Plate 10(b)
The north-west corner of the platform, showing the Chanter riser gantry. The gantry is the two structures, one ending in a triangular shape and the other in a rectangular one, projecting from the 68ft level. Also shown is the 20ft level and the navigation aid platform at the north-west corner of the 68ft level.

Plate 11(a)   The *Tharos*.

Plate 11(b)   The *Silver Pit*.

Plate 12(a)   An Atlantic 21 fast rescue craft (FRC).

Plate 12(b)   A Piper liferaft on its launching platform.

Plate 13   The east flare before the initial explosion. The photograph was taken by Mr Macdonald from the *Lowland Cavalier* some 1-2 minutes before the initial explosion.

Plate 14(a)

Plate 14(b)

The early fire in B Module: (a) Mr Miller's first photograph; and (b) his second photograph. The first photograph was taken from the *Tharos* some 15 seconds after the initial explosion. The next five photographs were taken in quick succession at intervals estimated as 2-3 seconds.

Plate 15(a)

Plate 15(b)

The early fire in B Module: (a) Mr Miller's third photograph; and (b) his fourth photograph.

Plate 16(a)

Plate 16(b)
The early fire in B Module: (a) Mr Miller's fifth photograph; and (b) his sixth photograph.

Plate 17(a)   The early fire in B Module: photograph taken by Mr Macdonald from the *Lowland Cavalier* at a time estimated as some 30 seconds after the initial explosion.

Plate 17(b)   The fires on Piper before riser rupture: Mr Miller's thirteenth photograph.

Plate 18(a)   The fires on Piper before riser rupture: Mr Miller's nineteenth photograph.

Plate 18(b)   The fires on Piper before riser rupture: photograph taken from the *Maersk Cutter*.

Plate 19(a)   The fires on Piper before riser rupture: photograph taken by Mr Ritchie from the *Lowland Cavalier*

Plate 19(b)   The fires on Piper just after rupture of the Tartan riser: photograph taken by Mr Gibson from the *Lowland Cavalier* at a time estimated as 22.20–22.22 hours.

Plate 20(a) The fires on Piper some time after riser rupture: photograph taken from the *Tharos* helicopter.

Plate 20(b) The fires on Piper some time after riser rupture: further photograph taken from the *Tharos* helicopter.

Plate 21 The remains of A Module on the morning of 7 July.

Plate 22(a)

Plate 22(b)

Accommodation modules at Flotta after recovery from the seabed — 1: (a) the East Replacement Quarters (ERQ); and (b) the south face of the ERQ.

Plate 23 Accommodation modules at Flotta after recovery from the seabed — 2: the Additional Accommodation West (AAW).

Plate 24(a)

Plate 24(b)

The pipework at the site of PSV 504 in C Module: (a) model of the PSV and pipework; and (b) a 1500 lb flange with ring. The PSV is the wooden mock-up. Its inlet is on the right hand side.

Plate 25(a) Hydrate formation in an observation window of the rotating wheel test rig in the tests conducted by Petreco.

Plate 25(b) The main methanol pump at Peterhead after recovery. As the pump was installed on the platform, the view is that looking north.

Plate 26(a) Explosion in a scaled down module typical of those conducted by CMI to validate its explosion model.

Plate 26(b) Leak from a blind flange assembly in the Nowsco tests.

Appendix A

Procedural History

Preliminary Hearings

A.1 A Preliminary Hearing was held at the Aberdeen Exhibition and Conference Centre, Bridge of Don, Aberdeen, on 11 November 1988. The date was selected in order to enable potential parties to have an adequate opportunity to consider the implications of The Petrie Report. At this hearing I disposed of applications by persons to be parties to the Inquiry and dealt with various matters of procedure and programming. I permitted persons to be parties to the Inquiry if they were able to show an interest, which required to be a reasonably direct interest, in some aspect of the subject matter of the Inquiry which as a matter of fairness required protection by such representation. I required that persons who had a similarity of interests should be jointly and not separately represented. Appendix B contains a list of the parties to the Inquiry.

A.2 Parties who intended to attach blame or criticism to someone else, whether or not already a party to the Inquiry, were required to give advance notice to the Secretariat which was then responsible for informing the other parties and any other person who had been so named. This procedure was carried out satisfactorily.

A.3 A second (and final) Preliminary Hearing was held in Edinburgh on 9 January 1989. This meeting was concerned solely with the progress which had been made in the recovery of documents by agreement.

The Inquiry

A.4 Part 1 of the Inquiry opened on 19 January 1989 and closed on 1 November 1989. It sat for 130 days. Part 2 of the Inquiry opened on 2 November 1989 and closed on 15 February 1990. It sat for 50 days. The whole proceedings of the Inquiry were held in public. No opening submissions were made. At the end of Part 1 closing submissions were made by the parties and Counsel to the Inquiry in writing, extending to over 1300 pages, and briefly highlighted orally. Copies of the written submissions were made available to the press and the public. At the end of Part 2 parties who intended to invite me to make recommendations submitted written lists of them in advance of making their closing submissions. Copies of these lists were likewise made available to the press and the public. During the course of the Inquiry a committee of experts under the chairmanship of the Assessors met on 2 occasions. The first of these was on 20 February 1989 and the second on 27 June 1989. These meetings were concerned respectively with (i) the involvement of the contents of pipelines with which Piper was connected; and (ii) expert opinion in regard to the scenario that condensate had been ingested into the reciprocating compressors.

Witnesses

A.5 196 witnesses gave evidence in Part 1 of the Inquiry. 14 of them who required to give evidence on a number of different matters gave evidence on 2 or more occasions. 64 witnesses gave evidence in Part 2 of the Inquiry. 9 of them gave evidence on 2 occasions. Lists of the witnesses are contained in Appendices C and E; and a list of experts' reports in Part 1 is contained in Appendix D. After I had decided that the evidence of a witness would be of assistance to the Inquiry copies of his precognition were circulated by the Crown Office to the parties. This served to avoid any risk of the Inquiry being delayed. Each witness gave evidence after being put on oath or affirming; and was subject to cross-examination by the parties to the Inquiry and, where applicable, by Counsel to the Inquiry, subject to the extent which I considered to be of assistance to the Inquiry. In certain instances I permitted legal representation to be provided in support of witnesses whose interests were not represented by any of the parties to the Inquiry.

A.6 At the Preliminary Hearing on 11 November 1988 the Lord Advocate (The Lord Cameron of Lochbroom) attended in the public interest and explained the assistance which he had arranged that the Crown Office should provide to the Inquiry. He also made the following statement:

"Further, to assist the Inquiry, I wish to state that I undertake that in the case of any witness who appears before this Inquiry, neither his evidence before the Inquiry nor evidence given before the Inquiry by any person by reference to or incorporating the whole or any part of any documents which that witness is required to produce to the Inquiry, shall be used against him, or any other person who could be held criminally responsible on account of his actings in any subsequent criminal proceedings, except in criminal proceedings in which the charge is one of having given false evidence before the Inquiry or having conspired with or procured others to do so."

Documents

A.7 A large number of documents were recovered for examination in order to see whether they might be of assistance to the Inquiry. The principal sources were Occidental, Score (UK) Ltd, the DEn and Lloyds Register of Shipping. The documents recovered from Occidental included a number of logs and other records which had been recovered from the bed of the North Sea when the ERQ was raised in October 1988. As the result of remarkable work which was done by specialists at Flotta virtually all of these documents were still legible. In order to receive the large volume of recoveries a library was set up and administered by Messrs Cremer and Warner in order that all parties might have access to them. From the contents of the library Messrs Cremer and Warner assembled a collection of core documents which form the first instalment of the productions before the Inquiry. These were copied and circulated to all the parties. Thereafter additional documents were produced from the library and elsewhere as the need for the Inquiry to refer to them was identified.

A.8 Due to the co-operation of the havers of the documents it proved unnecessary for me to exercise my power under Reg 7(a) of the Public Inquiries Regulations to require production of documents considered necessary for the purposes of the Inquiry. The sole exception to this was the interim report of the Occidental board of inquiry into the disaster which counsel for the Contractors' Interests asked me on Day 78 to require Occidental to produce. This motion was opposed by Occidental on a number of grounds. Having heard counsel I considered that the appropriate course of action in the first instance was to pronounce such an order for the limited purpose of enabling Counsel to the Inquiry to advise me as to any grounds which there might be upon which the Court might consider it necessary for the purposes of the Inquiry to examine the report. I pronounced that order on 13 June 1989. On 16 June I was advised by Counsel to the Inquiry that the report did not contain information which would add materially to what was already available to the Inquiry. In these circumstances I made no further order in regard to the interim report.

A.9 As regards all documents circulated to the parties by the Crown Office, such as witness statements and productions, the representatives of all parties entered into an agreement with Counsel to the Inquiry that these were provided solely for use by the parties and their agents, experts and representatives in connection with the preparation for and conduct of the proceedings; and gave various undertakings including not to publish their contents beyond the extent to which they had already been made public in the proceedings.

Costs and expenses

A.10 On 9 November 1989 I issued the following Opinion in connection with an application by the Trade Union Group for a direction as to their expenses:-

The Piper Alpha Trade Union Legal Group, which I shall refer to as "the Group", have applied to me for a direction that the expenses of the Group so far as properly attributable to its participation in Part 1 of the Inquiry should be paid by Occidental. This application, which is opposed by Occidental, was made in reliance on Reg 9(2) of The Offshore Installations (Public Inquiries) Regulations 1974, under which the Inquiry is held. That provides as follows:- "The court may direct that the costs of an inquiry shall be paid in whole or in part by any person who in the opinion of the court, by reason of any act or default on his part or on the part of any agent or servant of his, caused or contributed to the casualty or other accident the subject of the inquiry."

Written submissions in regard to this application have been presented to me by the Group and by Occidental. The Department of Energy and the Department of Transport have also submitted observations in writing. On Day 130 of the Inquiry I heard counsel for these parties and had the benefit of observations by Counsel to the Inquiry.

Counsel for the Group sought to satisfy me that the application was competent; and, if so, that it was made at an appropriate time. He further submitted that it was appropriate for me to make the direction in due course in the light of the view which he had earlier asked me to take as to Occidental's responsibility for what happened in the disaster.

The question of competency, upon which I have had full submissions from counsel, is logically the first point which requires to be addressed and it is appropriate that I should deal with it at the present stage. The short point is whether the expenses of a party such as the Group fall within the meaning of the expression "the costs of an inquiry".

Counsel for the Group submitted that when given its ordinary meaning, which involved eschewing "artificial refinement or subtle distinctions", the expression covered all the costs which were properly incurred by any party whose representation before the Inquiry had been permitted by the court. He pointed out that under Reg 4 the court is to hold the inquiry "in such a manner and in such conditions as the court thinks most effectual for enabling it to make the report required by Reg 9". One of the ways in which it did so, he said, was by scrutinising applications by persons to be represented as parties before the Inquiry. Where permission was given, the basis for that must have been that the court was satisfied that representation of that party before the Inquiry would be of assistance to it. He went on to submit that whereas the remuneration of the court and the assessors was specifically dealt with in Reg 3(2) it was clear that a number of items of expense which were not mentioned in the Regulations but must have been considered by the court to be of assistance to it, were covered by the expression "the costs of an inquiry". He gave as examples the cost of secretarial assistance, accommodation for the Inquiry and the services of Counsel to the Inquiry. By similar reasoning the expression would cover the expense incurred by a person who had been permitted to be a party to the Inquiry. In support of his arguments counsel founded on the decision of Lord Ross in *Holburnhead Salmon Fishing Co. v Scrabster Harbour Trustees*, 1982 SC 65. In that case, he held that the words "all costs, charges and expenses of and incidental to the preparing and obtaining of this Order and otherwise incurred in reference thereto shall be paid by the Trustees" in a provisional order which had been confirmed were wide enough to cover expenses incurred by the company in pursuing its petition against the order. Counsel drew attention to the fact that at page 66 Lord Ross described this language as very wide, adding that he could see no reason why the section should be construed in a narrow sense. Counsel also founded on the terms of certain orders made by the Hon Mr Justice Sheen as Wreck

Commissioner in the Zeebrugge Inquiry under section 56(5) of the Merchant Shipping Act, 1970, which provides that "The Wreck Commissioner or Sheriff may make such order with regard to the costs of the investigation as he thinks just ....". In a supplementary report on costs the Wreck Commissioner stated, *inter alia*, that "the statutory power of a Wreck Commissioner to make an order for costs is laid down in the most general terms". He went on to refer to the principle that an order for costs was used as the only method of penalising which was competent to the court as against parties other than certificated officers. The orders made by him in that inquiry included orders for various payments by Townsend Car Ferries Ltd to representatives of the National Union of Seamen and of certain members of the crew or their dependants, all as orders "with regard to the costs of the investigation".

The above arguments make it necessary for me to examine the true significance of the fact that a person has become a party to the present inquiry. It is to be noted that the 1974 Regulations do not compel, or empower the court to compel, any person to be a party to the inquiry. Further they do not require the court to accept any person as such a party. On the other hand the terms of Reg 7(b) and (d) refer to "any person appearing" at the inquiry and plainly imply that the court has the power to permit persons to appear as parties before the inquiry. That is a power which I exercised in the present inquiry at the Preliminary Hearing on 11 November 1988 by granting permission in the exercise of my discretion and on cause shown. The words in which I expressed what required to be shown, the substance of which would in any event have been implied, were that "I must be satisfied that the person has an interest, and that is a reasonably direct interest, in some aspect of the subject matter of the Inquiry which, as a matter of fairness, requires protection by such representation." Accordingly the basis for giving permission was the possession by a person of such an interest and not the expectation on the part of the court as to the assistance which was expected from him. It is not essential to an inquiry such as the present for any particular persons to be parties to it. In these circumstances I consider that the attempt to make a comparison between the cost of various items which have been provided at the request of the court or those acting on its behalf and the cost of a party's representation before the inquiry is ill-founded.

When the provisions of the 1974 Regulations are further considered a number of other important points emerge. At no point in the Regulations is there any explicit reference to expenses attributable to a person appearing before the inquiry or any mechanism by which a test is provided in order to ensure that the expenses were justified or reasonable. On the other hand express provision is made by Reg 8 for the payment and taxation of the expenses of witnesses. Turning again to Reg 9, para (3) provides that any costs which a person is ordered to pay under para (2) may be recovered from him by the Secretary of State. No provision is made for recovery by any other person, as one would have expected if "the costs of an inquiry" included the expenses attributable to a party's appearance before the inquiry. Para (5) provides that "The costs of an inquiry, other than any costs paid by any person pursuant to a direction of the court under para (2), shall be treated as expenses of the Secretary of State under the Act". There is nothing to suggest that in para (5) the meaning of "the costs of an inquiry" is different from the meaning of that expression where it appears in para (2). If the submissions of the Group are correct it means that, apart from any direction under para (2), the expenses of all parties to the inquiry which are attributable to their participation are to be treated as expenses of the Secretary of State under the Mineral Workings (Offshore Installations) Act 1971, and accordingly by virtue of section 13(1)(a) of that Act, to be paid out of money provided by Parliament. It should also be noted that this would be regardless of whether as matters turned out their representation before the inquiry, which could have involved little or no active participation, was of any assistance to the inquiry. Further, unlike the situation in para (2), there would be no question of any exercise by the court of a discretion which the Group submit can and should be exercised in their favour. These considerations reinforce the views which I have expressed above as to the unsoundness of the Group's submissions.

As regards the two examples of the approach to expenses cited by counsel for the Group I do not find them to be of any real assistance since they depend upon the statutory provisions to which each was related. In *Holburnhead Salmon Fishing Co. v Scrabster Harbour Trustees* it is clear that Lord Ross attached particular importance to the words "and otherwise incurred in reference thereto" which have no parallel in the present case. As regards the orders made at the Zeebrugge Inquiry, there does not appear to have been any dispute as to the competency of making such orders. Further, as counsel for the Department of Energy and the Department of Transport pointed out, it may well be of significance that in terms of the Merchant Shipping (Formal Investigations) Rules 1985, under which that Inquiry was held, certain persons could be made parties to the formal investigation by service of a notice of investigation upon them.

I should add that counsel for the Department of Energy and the Department of Transport also drew to my attention the way in which the Civil Aviation Review Board which was concerned with the Chinook helicopter accident at Sumburgh approached the question of expenses under Reg 14(7)(a) of the Civil Aviation (Investigation of Accidents) Regulations, 1983. In that case the Board, of which the Chairman was Sheriff P G B McNeill, QC, took the view that "the expenses of the review board" included the expenses of persons who were obliged by the Regulations to render assistance to the Review Board but did not include parties who appeared by virtue of Reg 14(2). While this decision is of some interest within its own context, I attach no importance to it for present purposes.

For the reasons which I have set out above I have come to the conclusion that the Group's attempt to bring their expenses within the scope of the expression "the costs of an inquiry" is ill-founded and accordingly I reject their application as incompetent.

A.11 At the end of the Inquiry counsel for the Trade Union Group invited me to make a recommendation on an extra-statutory basis that "payment of the costs incurred by ... MSF and the T & GWU should be made out of central funds." In support of this he pointed out that members of these unions were among the deceased and the survivors. The unions had sought to represent the interests of their members in seeking the reasons for the toll of deaths and injuries: and to promote safety in a number of specific areas of the safety regime, including by its alteration in a number of respects. He suggested that the Inquiry had been assisted by that representation, in particular in cross-examination of the Occidental management and those responsible for the operation of the safety regime. The unions had borne a heavy burden of expenditure. There were considerations of public policy in favour of not discouraging trade unions and other bodies and persons of limited resources from participation in public inquiries into matters of safety. He also drew my attention to recommendations which the inspector had made in the investigations into (i) the King's Cross Underground Fire (in favour of the Fire Brigade's Union and ASLEF); and (ii) the Clapham Junction Railway Accident (in favour of the NUR, TSSA and ASLEF). I have carefully considered these submissions. In my view these submissions have merit, particularly when it is borne in mind that, in regard to the opening up of matters of possible criticism, Counsel to the Inquiry adopted a neutral position. Accordingly the burden of exploring such matters fell to a large extent to the trade unions. On the other hand not all the points which they sought to explore proved to be of assistance to the Inquiry. While the matter of any payment out of public funds is entirely at the discretion of the Secretary of State I recommend that these trade unions should receive a contribution towards their costs. Having considered all the relevant factors I recommend that 40% would be an appropriate proportion. If the Secretary of State is disposed to make a payment out of public funds I recommend, in line with the course proposed by counsel for the Trade Union Group, that the costs should be taxed, failing agreement, by the Auditor of the Court of Session.

Appendix B

List of Parties and their Representatives

The Inquiry

The Solicitor-General for Scotland (Mr A F Rodger QC), Mr T C Dawson QC, Advocate-depute, Mr A P Campbell and Miss M Caldwell, Advocates; Mr A D Vannet, Solicitor, Crown Office, Edinburgh.

Piper Alpha Trade Union Legal Group (comprising 2 firms of solicitors for the representatives of 23 deceased and 8 survivors; and MSF and T & GWU).

Mr H H Campbell QC, Mr I Truscott, Advocate; Messrs Robin Thompson and Partners and Messrs Allan McDougall & Co, SSC, both of Edinburgh.

Piper Disaster Group (comprising 154 firms of solicitors for the representatives of 142 deceased and 49 survivors; and the EETPU).

Mr R N M MacLean QC, Mr C M Campbell, Advocate; Messrs Balfour and Manson, SSC, Edinburgh.

Occidental Petroleum (Caledonia) Ltd, Occidental Petroleum Corporation, Occidental Petroleum (Great Britain) Inc, Texaco Inc, Texaco Britain Ltd, Union Texas Petroleum Holdings Inc, Union Texas Petroleum Ltd and LASMO (TNS) Ltd

Mr J L Mitchell QC, Mr D G Monaghan and Mr D W Batchelor, Advocates; Messrs Paull and Williamsons, Advocates in Aberdeen.

Contractors' Interests (representing the interests of 25 offshore contractors - Aberdeen Offshore Services Ltd, Aberdeen Scaffolding Ltd, Bawden International Ltd, British Telecom PLC, Caleb Brett International Ltd, Eastman Christensen, Exploration Logging Ltd, W R Grace Ltd, Halliburton Manufacturing & Services Ltd, Inspectorate (UK) PLC, Kelvin Catering Ltd, Leuven Services (Aberdeen) Ltd, London Bridge Engineering Ltd, Macnamee Services Ltd, MI Great Britain Ltd, M B Services, McPherson Associates, Neyrfor UK Ltd, N L Petroleum Services (UK) Ltd, Northern Industrial & Marine Services Ltd, Orbit Valve Company Europe, Stena Offshore Ltd, Testwell Services Ltd, Wood Group Engineering Offshore Ltd, Wood Group Valve & Engineering Services Ltd).

The Dean of the Faculty of Advocates (Mr A C M Johnston QC), Mr R S Keen and Mr H W Currie, Advocates; Mr D M G Russell of Messrs Simpson and Marwick, WS, Edinburgh.

*Score (UK) Ltd**

Mr M S Jones QC; Messrs McClure Naismith Anderson & Gardiner, Solicitors, Glasgow.

*Ingersoll-Rand Co Ltd**

Mr G N H Emslie QC, Mr P Atherton, Barrister and Mr J R Campbell, Advocate; Messrs Lace Mawer, Solicitors, Manchester.

*Dresser Rand (UK) Ltd**

Mr D J D Macfadyen QC; Mr J R Foster QC (of the English Bar), Mr J G Reid, Advocate; Messrs Elliott & Co, Solicitors, Manchester.

*Allison Gas Turbine**

Mr F H Bartlit of Messrs Kirkland & Ellis, Attorneys, Chicago; Miss S Mason, Attorney; Messrs Dundas and Wilson, CS, Edinburgh.

Department of Energy and Department of Transport

Mr J M McGhie QC, Dr Lynda Clark QC; Mr A Williams of the Office of the Solicitor to the Secretary of State, Scottish Office, Edinburgh.

United Kingdom Offshore Operators Association Limited#

Mr A R Hardie QC; Messrs McGrigor Donald, Solicitors, Glasgow.

Notes:

The symbol * denotes that the representation was during Part 1 only of the Inquiry. The symbol # indicates that the representation was during Part 2 only.

In addition to the above the following were permitted to appear at the Inquiry:-

Miss Verity Jenner of Messrs Raeburn Christie & Co, Advocates in Aberdeen, on behalf of NUMAST, on day 56 for the interests of Mr J W Sabourn; and Mr N F Davidson, Advocate and Mr C R R Cowie of Macroberts, Solicitors, Edinburgh, on days 64, 65, 66 and 129 for the interests of Texaco North Sea (UK) Co Ltd.

Appendix C

List of Witnesses in Part 1

1.	AMAIRA, E Z	Diver (Survivor)
2.	ANSTOCK, C	Detective Inspector, Identification Bureau, Grampian Police
3.	ASHBY, A R	Former Deputy OIM, MSV Tharos
4.	ATKINSON, F H	Manager, Offshore Division, Lloyds Register
5.	BAGNALL, G H	Lead Maintenance Technician, Occidental
6.	BAKKE, J R DR	Manager, Explosion Research Laboratories, Christian Michelsen Institute, Norway
7.	BALFOUR, D DR	Director, Sieger Ltd
8.	BALLANTYNE, R H	Electrician (Survivor)
9.	BARCLAY, R W	Mechanical Fitter, formerly with Wood Group Valve and Engineering Services Ltd
10.	BARR, J	Diving Supervisor (Survivor)
11.	BARRON, W P	Chargehand/Foreman Painter (Survivor)
12.	BERRIFF, P	Independent Television Producer
13.	BETT, K E DR	Senior Lecturer, Department of Chemical Engineering, Imperial College, London
14.	BLAIR, D I	First Mate/DP Operator, MSV Tharos
15.	BODIE, A	Offshore Safety Superintendent, Occidental
16.	BOLLANDS, G	Production Operator (Survivor)
17.	BRADING, J E	Chairman, Occidental Petroleum (Caledonia) Ltd
18.	BRADLEY, M J	Rigger (Survivor)
19.	BRUCE, A C	Valve Technician, formerly with Score (UK) Ltd
20.	BURNS, J	Shift Supervisor - MCPO1, Total Oil
21.	BUSBY, F	Driller (Survivor)
22.	CALDER, H J	Helicopter Landing Officer (Survivor)
23.	CAREY, R F	Instrument Technician (Survivor)
24.	CARR, W H	Director, John Wood Group PLC and Wood Group (Engineering) Ltd
25.	CARROLL, A M	Inspection Diver (Survivor)
26.	CARSON, G	Second Engineer/Medic, MV Silver Pit
27.	CASSIDY, N G	Instrument Technician (Survivor)
28.	CLARK, A G	Maintenance Leadhand (Survivor)
29.	CLARK, M R	Chief Process Engineer, Occidental
30.	CLAYSON, P G	Former Safety Superintendent, Occidental
31.	CLAYTON, W F	Scaffolder (Survivor)
32.	CLEGG, M	Master, MV Lowland Cavalier
33.	COMMON, R M	Site Administrator (Survivor)
34.	CORMACK, E J	Police Constable, Grampian Police
35.	COTTER, J E	Production Operator, Occidental
36.	COX, R A DR	Consultant, formerly Chief Executive, Technica Ltd, Consulting Scientists and Engineers
37.	CRAIG, J A	Valve Technician (Survivor)

38.	CROSS, J H	Managing Director, RGIT Survival Centre Ltd, Aberdeen
39.	CUBBAGE, P A	Consultant Scientist, Cremer and Warner Ltd, Consulting Engineers and Scientists
40.	CUNNINGHAM, K	Diver (Survivor)
41.	DAVIDSON, J	Operations Superintendent - Claymore, Occidental
42.	DAVIE, R A	Senior Consultant, Yard Ltd
43.	DAVIES, M E DR	Managing Director, BMT Fluid Mechanics Ltd
44.	DIXON, J P	Painter, formerly of Wood Group Engineering
45.	DRYSDALE, D D DR	Lecturer, Unit of Fire Safety Engineering, University of Edinburgh
46.	DRYSDALE, J	Production Operator, Occidental
47.	DUGUID, I	Lead Roustabout (Survivor)
48.	DUTHIE, D	Detective Sergeant, Grampian Police
49.	ELLINGTON, D	Rigger (Survivor)
50.	ELLIOTT, D	Foreman Rigger (Survivor)
51.	ELLUL, I R	Consultant Engineer, Scientific Software-Intercomp (UK) Ltd
52.	ENNIS, S O	Master, MV Sandhaven
53.	FERGUSON, I	Mechanical Technician (Survivor)
54.	FLAWS, W J	Deck Foreman, MSV Tharos
55.	FOWLER, I N	Joiner (Survivor)
56.	GIBSON, R	Construction Engineer, Coflexip (UK) Ltd - MV Lowland Cavalier (Photographs)
57.	GOODWIN, B	Chargehand Rigger (Survivor)
58.	GORDON, I M	Chief Inspector, Grampian Police
59.	GORDON, R McG	Manager, Loss Prevention Department, Occidental
60.	GORDON, T D	First Mate/DP Operator, MSV Tharos
61.	GRANT, P M	Manager of Human Resources, Bawden International Ltd
62.	GRIEVE, E C	Production Operator (Survivor)
63.	GRIEVE, J H K DR	Head of the Department of Forensic Medicine, University of Aberdeen
64.	GRIFFITH, I L	Helicopter Pilot, British International Helicopters
65.	GROGAN, G E	Vice President Engineering, Occidental
66.	GUIOMAR, D	OIM -MCPO1, Total Oil
67.	GUTTERIDGE, J L	Toolpusher (Survivor)
68.	HAFFEY, C A	Seaman, MV Silver Pit
69.	HENDERSON, J S	Commandant, Offshore Fire Training Centre, Montrose
70.	HENDERSON, T A	Lead Operator, Occidental
71.	HENDRY, W T DR	Former Head of the Department of Forensic Medicine, University of Aberdeen
72.	HILL, D J	Crane Operator (Survivor)
73.	HODGSON, S A	Flight Lieutenant - 202 Squadron, RAF Lossiemouth
74.	HUTCHISON, E	Nautical Surveyor, Maritime Directorate, Department of Transport
75.	JACKSON, B	Rigger (Survivor)

76.	JEFFREY, P G	Consultant Engineer, Plessey Assessment Services Ltd
77.	JENKINS, R D	Senior Inspector, PED, Safety Inspectorate, Department of Energy
78.	JENNINGS, M H G	Flight Information Logistics Officer (Survivor)
79.	JOHNSEN, H K DR	Managing Director, Petreco A/S, Norway
80.	JONES, J M CAPT	Managing Director, London Offshore Consultants
81.	KERR, E R	Radio Officer, British Telecom International, Wick Radio Station
82.	KHAN, M R	Chemist (Survivor)
83.	KILOH, A J	Deckhand, MV Silver Pit
84.	KINRADE, D H	Radio Operator (Survivor)
85.	KONDOL, J M	Deputy OIM, MSV Tharos
86.	LAMB, C W	Mechanical Fitter (Survivor)
87.	LAMBERT, D	Scaffolder (Survivor)
88.	LEEMING, J	Former OIM - Tartan, Texaco
89.	LETHAM, I	Former Deck Hand, MV Sandhaven
90.	LETTY, A D MacK	OIM, MSV Tharos
91.	LLOYD, P	Senior Electrical Engineer, Occidental
92.	LOBBAN, W J	Water Blaster (Survivor)
93.	LOCKWOOD, C	Lead Production Operator, Occidental
94.	LYNCH, J	Lead Production Operator, Occidental
95.	MACALLAN, J L	Production and Pipeline Manager, Occidental
96.	MCDONALD, A G	Head of Telecommunications - North Sea, Occidental
97.	MCDONALD, J McG	Rigger (Survivor)
98.	MACDONALD, L M T	Electronic Technician, UDI Group Ltd -MV Lowland Cavalier (Photographs)
99.	MCGEOUGH, F	Safety Training Co-Ordinator, Occidental
100.	MCGREGOR, R J	Mechanical Technician (Survivor)
101.	MACKAY, A J	Electrician (Survivor)
102.	MACKAY, I F	First Mate, MV Lowland Cavalier
103.	MCLAREN, W McI	Electrical Engineer Surveyor, Lloyds Register
104.	MACLEAN,R MacK C	Master, MV Loch Carron
105.	MACLEOD, S R	Diving Superintendent (Survivor)
106.	MCNEIL, D A DR	Senior Scientific Officer, National Engineering Laboratory
107.	MCNEILL, J P	Former Coxswain, MV Silver Pit
108.	MACPHERSON, C A	Master, MV Loch Shuna
109.	MCREYNOLDS, A D	Vice President - Operations, Occidental
110.	MARSHALL,J G DR	Consulting Scientist, formerly with Burgoyne and Partners
111.	MAY, D J McD	Senior Engineer for Pipelines and Structures, Marine Department, Occidental
112.	MEANEN, J S	Scaffolder (Survivor)
113.	MENZIES,J A R H	Scaffolder (Survivor)
114.	MIDDLETON, A H	Noise and Vibration Consultant/Director, Anthony Best Dynamics Ltd
115.	MIDDLETON, S J	Diver (Survivor)

116. MILLAR, A J	General Secretary, Professional DiversAssociation
117. MILLER, C A	Mobile Diving Unit Pilot, Aberdeen Offshore Services - MSV Tharos (Photographs)
118. MILLER, D A	Security Manager, Occidental
119. MITCHINSON, W	Former Mate, MV Silver Pit
120. MOCHAN, A H	Superintendent Engineer (Survivor)
121. MORETON, M D	Production Supervisor - Tartan, Texaco
122. MORTON, C I	Master, MV Maersk Cutter
123. MUIR, I F	Former Second Mate, MV Loch Shuna
124. MURPHY, P J	First Engineer, MSV Tharos
125. MURRAY, J	Helicopter Engineer, British International Helicopters
126. MURRAY, J	Production Operator, Occidental
127. NAYLOR, D E	Driller (Survivor)
128. NIVEN, C I	Diver (Survivor)
129. PALMER, A C DR	Managing Director, Andrew Palmer andAssociates Ltd, Consulting Engineers
130. PARRYDAVIES, G P	Diver (Survivor)
131. PATERSON, E A MRS	Former Process Chemical Engineer, Occidental
132. PATERSON, R E	Welder (Survivor)
133. PATERSON, W N	Chief Engineer, MSV Tharos
134. PATIENCE, J A	Lead Safety Operator, Occidental
135. PETRIE, J R	Director of Safety, PED, Safety Directorate, Department of Energy
136. PILLANS, G P	Senior Electrical Surveyor, Lloyds Register
137. PIRIE, A	Service Engineer, Wood Group Valve and Engineering Services Ltd
138. PLUMMER, C D	Chief Engineer, Atkins Oil And Gas Engineering Ltd
139. POUNTNEY, R J	Winchman - 202 Squadron, RAF Lossiemouth
140. POWELL, A C M	Crane Operator (Survivor)
141. PUNCHARD, E T R	Diving Inspection Controller (Survivor)
142. RAE, S	Electrician (Survivor)
143. RALPH, N E	Foreman Rigger (Survivor)
144. RANKIN, A D	Valve Technician (Survivor)
145. REID, M A	Lead Foreman (Survivor)
146. RICHARDS, G	OIM(Back to Back) - Piper Alpha, Occidental
147. RICHARDSON, S M DR	Senior Lecturer, Department of Chemical Engineering and Chemical Technology, Imperial College, London
148. RITCHIE, A	Detective Superintendent, Grampian Police
149. RITCHIE, A A	Civil Engineer, Ritchie Sub-sea Engineering - MV Lowland Cavalier (Photographs)
150. RITCHIE, C B	Managing Director, Score (UK) Ltd
151. ROBERTS, G D	Squadron Leader, RAF - Rescue Co-ordination Centre, Edinburgh
152. ROBERTS, K	Facilities Engineer - Tartan, Texaco
153. ROBERTSON, G G	Safety Supervisor, Occidental

412

154. ROBINSON, D T	Barge Clerk/Helicopter Landing Officer, MSV Tharos
155. ROGERS, T	Facilities Engineer, Occidental
156. ROWAN, C J	Senior Diving Superintendent, StenaOffshore
157. RUSSELL, J B	Mechanical Fitter (Survivor)
158. RUTHERFORD, J	Rigger, Wood Group Engineering Offshore Ltd
159. SABOURN, J W	Former Master, MV Silver Pit
160. SANDLIN, S B	OIM - Claymore, Occidental
161. SAVILLE, G DR	Department of Chemical Engineering and Chemical Technology, Imperial College,London
162. SCANLON, T J	Offshore Superintendent, formerly with Wood Group Engineering Offshore Ltd
163. SCILLY, N F DR	Principal Specialist Inspector, Technology Division, Health and Safety Executive
164. SCOTHERN, E	Instrument Technician, Occidental
165. SEDDON, R H	Senior Maintenance Superintendent, Occidental
166. SKIDMORE, M	Senior Facilities Engineer, Occidental
167. SLAYMAKER, J M	Production Operator, formerly with MB Services
168. SMYLLIE, R J	Senior Engineer, Cremer and Warner Ltd, Consulting Engineers and Scientists
169. SNEDDON, R G	Operations Superintendent, Occidental
170. STANDEN, R	Senior Physicist, NOWSCO Well Services Ltd
171. STICKNEY, M E DR	Senior Systems Engineer, Hughes Aircraft Corporation
172. STOCKAN, L W	Lead Process Operator, Flotta Terminal, Occidental
173. STRACHAN, R DR	Consultant, Aberdeen Industrial Doctors
174. STREET, W R	Director, Hollobone Hibbert and Associates Ltd
175. SWALES, V	Derrickman (Survivor)
176. SYLVESTER-EVANS, R	Associate Director, Cremer and Warner Ltd, Consulting Engineers and Scientists
177. TAIT, J	Service Engineer, Score (UK) Ltd
178. TEA, D C	Instrument Technician, Occidental
179. THOMPSON, D McR	Rigger (Survivor)
180. THOMSON, M S	Senior Engineer Surveyor, Lloyds Register
181. THOMSON, R G	Industrial Cleaner (Survivor)
182. THORNTON, P G	Assistant Firemaster, Grampian Fire Brigade
183. TODD, A C B	Maintenance Superintendent, Occidental
184. TUCKER, D M	Fire and Loss Consultant/Senior Partner, Tucker Robinson, Consulting Scientists
185. TURNER, D J	Managing Director, Camera Alive Ltd
186. WATT, A	Valve Technician, Score (UK) Ltd
187. WATTS, P C A	Chief Process Engineer, Kaldair Ltd
188. WELLS, J V	Diver (Survivor)
189. WHALLEY, D	Team Leader, Score (UK) Ltd
190. WILLIAMSON, R	Service Technician, Bran and Leubbe (GB) Ltd, Pump Manufacturers
191. WOOD, A L	Fitter (Survivor)
192. WOOD, J O	Diving Technician (Survivor)

193. WOOD, W P	Ship Surveyor, Marine Directorate, Department of Transport
194. WOTTGE, K R	Facilities Engineering Manager, Occidental
195. WYNN, J P A	District Staff Officer, HM Coastguard, Aberdeen
196. YOUNG, W H	Instrument Technician (Survivor)

(Note: Three witnesses, MACKENZIE, I H - Scaffolder (Survivor), MILLER, H J - Rigger (Survivor) and PAYNE, A G - Diver (Survivor), did not give evidence at the Inquiry but their statements were read out by Counsel to the Inquiry at the end of Part 1.)

Appendix D

List of Reports by Experts Submitted in Evidence in Part 1

1. **Andrew Palmer and Associates Limited**

   "Damage to 4 inch condensate line in Module B of Piper Alpha Platform" - May 1989 - spoken to by Dr A Palmer

2. **Anthony Best Dynamics Limited**

   "Investigation of noises heard prior to the first explosion" - September 1989 - spoken to by Mr A H Middleton

3. **Atkins Oil and Gas Engineering Limited**

   "Study of Piper Alpha liquid carry-over to compressors scenario" - June 1989 - spoken to by Mr C D Plummer

4. **BMT Fluid Mechanics Limited**

   "Airflow and gas dispersion study" - August 1989 - spoken to by Dr M E Davies

5. **Bran and Leubbe**

   "Investigation of the stroke settings of the six-head methanol injection pump" - September 1989 - spoken to by Mr R Williamson

6. **Camera Alive Limited**

   "Report on analysis of polaroid photograph of flange" - August 1989 - spoken to by Mr D J Turner

7. **Christian Michelsen Institute**

   7.1 "Simulation of gas explosions in Module C, Piper Alpha" - November 1988 - spoken to by Dr J R Bakke

   7.2 "Gas explosion simulation in Piper Alpha Module C using FLACS" - September 1989 - spoken to by Dr J R Bakke

8. **Cremer and Warner**

   8.1 "Preliminary review of potential causation scenarios" - January 1989

   8.2 "Possible explanations for the prolonged flaring and venting on the Piper Alpha platform after the initial explosion" - June 1989 - spoken to by Mr R Smylie

   8.3 "Report on scenarios put forward by the Department of Energy from scenario C" - August 1989 - spoken to by Dr R Sylvester-Evans

   8.4 "Review of evidence relating to the initial explosion on the Piper Alpha platform on 6 July 1988" - spoken to by Dr P Cubbage

9. **Department of Chemical Engineering and Chemical Technology, Imperial College**

   9.1 "An appreciation of the operation of the reciprocating compressors used on the Piper Alpha oil platform" - June 1989 - spoken to by Dr K E Bett

   9.2 "Hydrate formation on Piper Alpha" - June 1989 - spoken to by Dr S M Richardson

9.3 "Analysis of flows in Piper Alpha gas import and export lines" - June 1989 - spoken to by Dr S M Richardson

9.4 "Autoignition in line to site of PSV 504 on Piper Alpha" - August 1989 - spoken to by Dr S M Richardson

9.5 "Gas and liquid leakage from line 2-P-524-4″-F15" - September 1989 - spoken to by Dr S M Richardson

9.6 "Repressurisation of lines associated with condensate injection pump 2-G-200A on Piper Alpha" - June 1989 - spoken to by Dr G Saville

9.7 "Leakage from line 2-P-524-4″-F15 to site of PSV 504 on Piper Alpha" - August 1989 - spoken to by Dr G Saville

9.8 "Hydrate/ice formation and occurrence of low temperatures in flare, vent and recycle lines on Piper Alpha" - September 1989 - spoken to by Dr G Saville

9.9 "Hydrate formation downstream of Joule-Thomson valve PCV 721" - September 1989 - spoken to by Dr G Saville

9.10 "Overpressurisation of condensate injection pump 2-G-200B and associated lines on Piper Alpha" - September 1989 - spoken to by Dr G Saville

9.11 "Failure pressures in discharge and safety valve lines from pump 2-G-200B" - September 1989 - spoken to by Dr G Saville

10. Department of Fire Safety Engineering, University of Edinburgh

"Review of evidence relating to the development of the fire which followed the initial explosion in module C" - August 1989 - spoken to by Dr D Drysdale

11. Department of Forensic Medicine, University of Aberdeen

"Statement by Head of Department of Forensic Medicine, University of Aberdeen" - August 1989 - spoken to by Dr W T Hendry

12. Dr J G Marshall

"The Piper Alpha disaster - a preliminary report on the potential sources of ignition" - spoken to by Dr J G Marshall

13. Health and Safety Executive

13.1 "An assessment of the explosion effects, and conditions likely to give rise to such effects, on the Piper Alpha production platform" - August 1988 - spoken to by Dr N F Scilly

13.2 "An assessment of the explosion effects, and conditions likely to give rise to such effects, on the Piper Alpha production platform - Supplementary report" - May 1989 - spoken to by Dr N F Scilly

14. Hollobone Hibbert and Associates Limited

"Feasibility of the recovery of subsea wreckage from Piper Alpha" - January 1989 - spoken to by Mr W R Street

15. Hughes Training and Support Systems Group

"Report of photograph processing and enhancement" - July 1989 - spoken to by Dr M F Stickney Jr

16. Kaldair

"Technical study of the flare performance during the Piper Alpha incident" - February 1989 - spoken to by Mr P C A Watts

17. **London Offshore Consultants**

"Preliminary report on the adequacy of the investigation and subsequent report by Hollobone Hibbert and Associates Ltd entitled "Feasibility of the recovery of subsea wreckage from Piper Alpha" " - February 1989 - spoken to by Captain J M Jones

18. **National Engineering Laboratory**

18.1 "Leakage evaluation tests on a weld-neck flange and blind flange assembly in connection with the Piper Alpha Public Inquiry investigation" - July 1989 - spoken to by Mr R A Davie

18.2 "Assessment of the NEL flange leakage experimental results taken in connection with the Piper Alpha Inquiry" - September 1989 - spoken to by Dr D A McNeil

19. **Nowsco Well Service (UK) Limited**

"Leakage and related effects from a pipe under pressure" - spoken to by Mr F Standen

20. **Petreco**

"Investigation into hydrate properties and their possible formation within the Piper Alpha production process" - September 1989 - spoken to by Dr H K Johnsen

21. **Plessey Assessment Services**

"Piper Alpha liferaft investigation" - April 1989 - spoken to by Mr P G Jeffrey

22. **Scientific Software-Intercomp**

"Hydraulic study of pipelines associated with the Occidental Piper Alpha platform" - December 1988 - spoken to by Dr I Ellul

23. **Sieger Limited**

"Report on gas detection" - spoken to by Dr D Balfour

24. **Technica, Consulting Scientists and Engineers**

24.1 "Investigation of blast resistance of firewalls" - August 1988 - spoken to by Dr R A Cox

24.2 "Investigation of blast resistance of firewalls - Supplementary Report" - May 1989 - spoken to by Dr R A Cox

24.3 "Extent of damage caused by the initial explosion and probable effects on critical systems" - June 1989 - spoken to by Dr R A Cox

24.4 "Projectile effects of firewall disintegration" - June 1989 - spoken to by Dr R A Cox

24.5 "Investigation of failure times of Tartan gas riser due to varying heat loads" - June 1989 - spoken to by Dr R A Cox

25. **Tucker Robinson**

"Report on the examination of the fire and smoke damage in Piper Alpha accommodation modules, east replacement quarters and additional accommodation west" - June 1989 - spoken to by Mr D M Tucker

Appendix E

List of Witnesses in Part 2

(and the subject matter of their evidence)

1. ADAMS, A J

    Principal Pipeline Inspector, Safety Directorate of PED, Department of Energy
    > Pipeline Isolation Systems including Subsea Valves

2. ALLEN, C S

    Head of Alwyn Safety, Total Oil Marine PLC
    > Application of Computers to Permit to Work Systems for Offshore Installations

3. ASHWORTH, M

    Senior Control Engineer, BP International
    > Process Control and Emergency Shutdown Systems

4. BANKS, R P

    Supervisor of Engineering Design and Construction, Chevron (UK) Ltd
    > The Qualifications and Qualities required in an Offshore Maintenance Supervisor

5. BAXENDINE, M R

    Offshore Installation Manager, Shell (UK) Exploration & Production Ltd
    > Command Structure in an Emergency

6. BOOTH, M J

    Head of Operations Safety, Shell (UK) Exploration & Production Ltd
    > Escape Routes to the Survival Craft and the Helideck on Offshore Installations

7. BRANDIE, E F

    Safety and Compliance Manager, Chevron (UK) Ltd, Chairman of UKOOA Fire Protection Working Group, representative of CBI on OIAC
    > Factors for Enhancing the Integrity of Offshore Safe Haven Areas. An Alternative to Standard Firewater System Designs for UK Sector Offshore Installations

8. BROADRIBB, M P

    Central Safety Engineering Superintendent, BP Exploration
    > Subsea Isolation Valves - The BP Approach

9. CHAMBERLAIN, G A DR

    Technical Leader of Explosion Protection Review Task Force, Shell (UK) Exploration & Production Ltd
    > The Nature and Mitigation of Vapour Cloud Explosions

10. COX, R A DR

    Consulting Engineer, formerly Chief Executive of Technica Ltd
    > Overview of Quantified Risk Assessment

11. CUNNINGHAM, A W T

    Occupational Health and Safety Officer, EETPU
    > Safety Representatives

12. DALZELL, G A

    Fire and Safety Engineer, BP International, Member of UKOOA Fire Protection Working Group
    > The Prevention of Smoke Ingress into Offshore Accommodation Modules

13.	DANIEL, J J S	Director, Hollobone Hibbert & Associates Ltd, Chairman of the Standby Ship Operators Association Ltd Standby Vessels
14.	DAVIES, G H	Area Director, Health and Safety Executive, Merseyside & Cheshire Area Permits to Work
15.	DAY, J	Head of Electrical Engineering, Shell Exploration & Production Ltd, Member of UKOOA Electrical Sub-committee Electrical Power for Emergency Systems
16.	DE LA PENA, M	Divisional Director, Environmental & Safety Products Division, Dowty PLC Smoke Hoods
17.	DENTON, A A DR	Chairman, Noble Denton International Ltd, Vice President of the Institution of Mechanical Engineers Quality Management Systems
18.	DOBLE, P A C	Deputy Project Manager, Kittiwake Project, Shell (UK) Exploration & Production Ltd The Means of Preventing and Mitigating the Effects of an Explosion - Kittiwake Project
19.	DREW, B C	Chief Surveyor, Marine Directorate, Department of Transport Code for Assessment of the Suitability of Standby Vessels
20.	ELLICE, K A J	Training Manager, BP Exploration Training of Offshore Installation Managers
21.	ELLIS, A F DR	Deputy Chief Inspector, Technology Division, Health & Safety Executive Quantified Risk Assessment - HSE's View
22.	EVANS, J D	Research & Development Manager, MSA (Britain) Ltd Smoke Hoods
23.	FERROW, M	Manager, Safety & Quality Assurance, Conoco (UK) Ltd The Offshore Safety Regime and Formal Safety Assessments
24.	FLEISHMAN, A B	Senior Safety Technologist, Group Safety Centre, BP International Gyda Safety Evaluation
25.	GILBERT, R B DR	Chief Engineer, Nelson Project Team, Shell (UK) Exploration & Production Ltd Subsea Valves
26.	GINN, M C CAPT	Principal Air Transport Officer, British Gas PLC, Chairman of UKOOA Aircraft Committee The use of Helicopters in Offshore Evacuation
27.	GORSE, E J	Principal Inspector, Safety Directorate of PED, Department of Energy Formal Safety Assessments
28.	HEIBERG-ANDERSON, G	Platform Manager, Gullfaks C, Statoil, Norway The Means of Ensuring Safe and Full Evacuation - The Statoil Approach, The Control of the Process

29.	HIGGS, F	National Secretary of the Chemical, Oil & Rubber Group, Transport & General Workers' Union The Offshore Safety Regime
30.	HODGKINS, D J	Director of Safety and General Policy Division, Health and Safety Executive The Agency Agreement between the Health and Safety Commission and the Department of Energy
31.	HOGH, M S DR	Manager Projects and External Affairs, Group Safety Centre, BP International Overview of Use and Value of Quantified Risk Assessment
32.	JONES, M J	Training Officer, Central Training Division, BP Exploration Development of Craft Training Scheme
33.	KEENAN, J M	Assistant General Secretary, Banking Insurance & Finance Union, formerly District Officer, Transport & General Workers' Union, Aberdeen Standby Vessels
34.	KELLEHER, T W	Fire and Safety Engineer, Shell Exploration & Production Ltd, Project Manager of Department of Energy/UKOOA Research Projects on Evacuation by TEMPSC Survival Craft and Free Fall Lifeboats
35.	KINLOCH, D S	Offshore Installation Manager, Conoco (UK) Ltd Independent Actions during Emergencies, Permit to Work Procedure
36.	KYLE, S R	Environment and Safety Co-ordinator, Brae Operations, Marathon Oil (UK) Ltd, Chairman of UKOOA Working Group on Permits to Work Permit to Work Procedure
37.	LIEN, E	Technical Director, Selantic Industrier as, Norway Skyscape Offshore Emergency Evacuation System
38.	LITTLEJOHN, I J	Process and Maintenance Engineering Group Supervisor, Amoco (UK) Exploration Ltd The Qualifications and Qualities Required in an Offshore Supervisor
39.	LYONS, R A	Assistant General Secretary, Manufacturing, Science & Finance Union The Offshore Safety Regime
40.	MCINTOSH, A R	Principal Inspector, Safety Directorate of PED, Department of Energy Protection against Fire and Explosion
41.	MCKEE, R E	Chairman and Managing Director, Conoco (UK) Ltd Managing Safety
42.	MACEY, M	Director, Maritime Rescue Services Ltd Standby Vessels - Training of Personnel
43.	MATHESON, A B MR	Consultant in Accident and Emergency Care, Aberdeen Royal Infirmary The Offshore Specialist Team
44.	MARSHALL, V C DR	Chartered Engineer, Formerly Director of Safety Services, Bradford University Safety Cases and Safety Assessments

45.	MIDDLETON, C J	Marine Superintendent, Marathon (UK) Ltd, Chairman of UKOOA Marine Committee Standby Vessels
46.	NORDGARD, T	Vice President, Projects Division, Statoil, Norway The Location and Protection of Accommodation on Integrated Drilling and Production Platforms on the Norwegian Continental Shelf
47.	OGNEDAL, M	Director, Safety and Working Environment Division, Norwegian Petroleum Directorate The Norwegian Offshore Safety Regime
48.	PAPE, R P DR	Head of Major Hazards Assessment Unit, Health and Safety Executive Quantified Risk Assessment - HSE's Experience
49.	PERROTT, I R	Assistant Chartering Manager, Maersk Co Ltd Skyscape
50.	PETRIE, J R	Director of Safety, PED, Department of Energy Life-Saving Appliances, The Offshore Safety Regime
51.	PRIDDLE, R J	Deputy Secretary, Department of Energy The Offshore Safety Regime
52.	RIMINGTON, J D	Director General, Health and Safety Executive The Onshore Safety Regime
53.	RUDD, D T	Marine Superintendent, BP Exploration Evacuation Policy and Plan for Forties Field
54.	SCANLON, T J	Mechanical Piping Engineer (formerly Offshore Superintendent, Wood Group Engineering Offshore Ltd) Permit to Work Systems
55.	SEFTON, A D DR	Leader of the Hazardous Installations and Transport of Dangerous Substances National Interest Group, Health and Safety Executive The Control of Industrial Major Accident Hazards Regulations 1984, Safety Reports
56.	SHEPPARD, R A	Vice President (Production) and Director, Amoco (UK) Exploration Co Ltd Managing Safety
57.	SIDE, J DR	Senior Policy Scientist, Institute of Offshore Engineering, Heriot-Watt University Offshore Emergency Rescue and Evacuation
58.	SPOUGE, J R	Consulting Senior Engineer, Technica Ltd Comparative Safety Evaluation, Arrangements for Accommodating Personnel Offshore
59.	TAYLOR, B G S DR	Director of Technical Affairs, UKOOA The Development and Future of the Offshore Oil Industry
60.	TVEIT, O J	Senior Engineer, Statoil, Norway Risk Assessment, The Norwegian Offshore Safety Regime
61.	VAN BEEK, A W	Head of Offshore Structures Engineering, Shell (UK) Exploration & Production Ltd Blast Walls

62. VASEY, M W DR   Manager, Safety Modelling and Offshore Safety, British Gas PLC Midlands Research Station
         Possible Mitigation of Module Explosions on Offshore Platforms

63. WALLACE, I G   Superintendent of Occupational Safety and Health, Conoco (UK) Ltd, Member of UKOOA Safety Committee and Chairman of Department of Energy Emergency Evacuation Steering Committee
         Emergency Evacuation and Escape/TEMPSC, Methods of Emergency Escape to Sea

64. WILLATT, R   Senior Pipeline Engineer, BP Engineering Piplines Group
         Functional and Safety Aspects of Offshore Pipeline Connections

424

Appendix F

Supplementary Material on Chapter 3

F.1 A description of the Piper Alpha platform was given in Chapter 3. Further information on certain detailed features is given here.

Centrifugal compressors

F.2 There were 3 parallel centrifugal compressor trains. The A train consisted of a suction scrubber, a centrifugal compressor, 1-K-105A, a gas cooler, and a discharge scrubber; the B and C trains were similar. Each compressor together with its turbine was housed in its own separate compartment at the extreme east end of C Module and the associated equipment was located on the centrifugal compressor gas skid inboard of the compressors themselves.

F.3 The function of the suction scrubber was to remove any condensate droplets not removed in the condensate knockout drum and carried forward. In normal operation there would be virtually no condensate removed at this point. Condensate would be formed, however, following compression and cooling and this condensate was removed in the discharge scrubbers. There was a level controller on each discharge scrubber.

F.4 Each compressor was driven by its own gas turbine. Air entering the first section, the gas generator, was compressed, fuel gas was then injected and the resultant mixture burnt and then expanded through 2 sets of turbines, the first to drive the air compressor just referred to, and the second, the power turbines, to drive the centrifugal compressors. The exhaust gases from these turbines were vented through tall exhausts at the east face of the module. The turbines were supplied with fuel gas from the fuel gas system. The fuel gas line within the turbine compartment included a hose section.

F.5 The compressor trains were equipped with gas operated valves (GOVs) to allow them to be shut in. There were a considerable number of trips on the turbines or the compressors themselves, including high gas discharge temperature; high suction and discharge scrubber levels; high and low fuel gas pressure; enclosure high temperature, fire and gas (50% LEL); seal and lube oil systems; and vibration. There was a seal oil system on each compressor to prevent gas escaping. On shutdown of the compressor the seal oil system would also shut down, though after a time delay. If the compressor was still pressurised, gas could escape and therefore on shutdown the compressor was automatically vented. There were recycle loops on the compressors and anti-surge controls to maintain the flow of gas through the machines by recycling and thus preventing them going into surge conditions.

F.6 Each compressor set was housed in an individual enclosure, made of steel sheet, the gas turbine and the compressor being in separate compartments of the enclosure, separated by a bulkhead, with the turbines outboard. The turbine compartment had double doors on the south side and the compressor compartment a single door also on the south side. The controls for the compressor, and its turbine, were on a local control panel, which was situated on the west side of the enclosure.

F.7 The turbine air intakes and exhausts and the enclosure ventilation are shown in Fig F.1. Air was drawn in to the turbine intakes through filter-silencer units with inlets located on both north and south sides at the east end of the turbine compartment. Burnt gas passed out through the exhausts, the outlets of which were high up, facing east, on the east side. Air for the ventilation of both turbine and compressor compartments was taken in at a south-facing intake at the east end of the enclosure. The source of this air was outside the module and from a safe area. Ventilation air from the turbine compartment passed out through a duct which terminated on the east side, while air from the compressor compartment passed out into the module through louvres on that compartment.

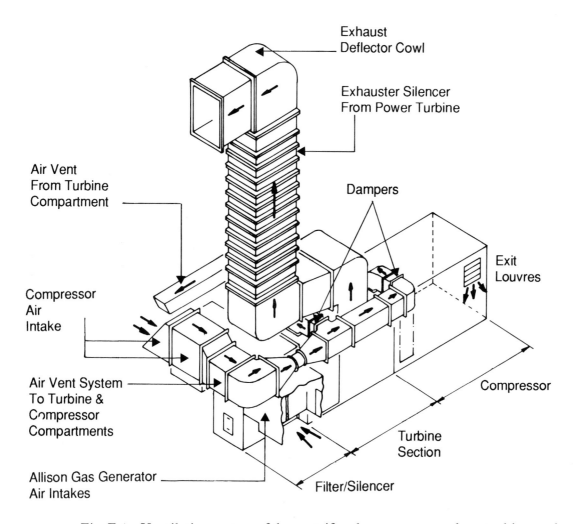

Fig. F.1 Ventilation system of the centrifugal compressor and gas turbine enclosures.

F.8 The ventilation system was designed to trip if the compressors stopped, but it was fitted with a time delay relay which permitted the ventilation to continue running for 2 hours after the turbine and compressor were shut down. However, if the gas detector at the ventilation air inlet registered a high gas alarm, both the compressor and ventilation systems would shut down immediately. The setting of this alarm was said to be 50% LEL, as for the turbine and compressor compartments, but this was not confirmed.

F.9 The compressor enclosures were about 10 ft high and stood on the solid deck of the module. There was a grating 2 ft above the deck and around the compressor set enclosure. About 5 ft to the west of the centrifugal compressor enclosure the solid deck sloped up and joined the grating. It then ran at that level until it reached the reciprocating compressors, after which it dropped 2 ft again and continued thus to the west side. There was no connection through which gas could flow at deck plate level from the east to the west side, because of the rise part way through the module. There was a half-door, starting at 3 ft and ending at 10 ft up, between the compressor enclosure and firewall on compressor C and between the 2 enclosures of C and B compressors, at the east end of the turbine filter-silencers and another door at the west end of these; the arrangements between the other compressors were not explored.

Reciprocating compressors

F.10 There were 2 parallel trains of reciprocating compressors with first and second stage compression. The first stage of the A train consisted of a suction scrubber, a reciprocating compressor, 1-K-103A and a gas cooler. The second stage of the A train consisted of a suction scrubber, and a reciprocating compressor, 1-K-103A (again); there was no gas cooler. The B train was similar. These compressors were located in the middle of C Module.

F.11 Following first stage compression the gas followed different paths, depending on whether the operational mode was phase 1 or phase 2, but in both cases it then entered the suction of the second stage.

F.12 The 2 stages of compression in each train were performed by a single machine. These were large machines: the motor and machine together were said to be about 35 ft long and the machine itself weighed some 70 tons. The associated equipment was located around the compressors. Each compressor was driven by an electric motor. The compressor consisted of 6 cylinders: 3 on the first stage and 3 on the second; the cylinders were double-acting. Each machine was oriented with its frame end to the south.

F.13 On each compressor train there were GOVs to allow it to be shut in and a number of trips which would operate to shut down and isolate the machine. There was a recycle loop around the first stage of each compressor and 2 recycle loops around the second stage, one through GOVs 903 and 905 and one through PCV 746. Some of the cylinders on each compressor were fitted with an unloader, a device which holds open one of the valves and thus prevents compression occurring. On the first stage the ends of the cylinder near the frame could not be unloaded but the outboard ends could. On the second stage 2 of the 3 cylinders could be unloaded on the outboard ends, but 1 cylinder could not be unloaded at all. The ability to unload and recycle gas around the compressors gave the flexibility to operate at low gas flows and to reduce the flow of gas going forward and the condensate produced. The controls for unloading and recycling were beside the machines in C Module. Unloading a compressor and putting it on recycle involved switching 7 switches: 3 unloaders and a recycle valve for the first stage, 2 unloaders and a recycle valve for the second.

F.14 In phase 2 operation the first stage reciprocating compressor capacity could be boosted by the use of the reciprocating compressor, 4-K-803, the SEPCO, or Worthington, compressor, located in the GCM, which was operated in parallel with the other 2 machines. Gas was taken to this compressor from the centrifugal compressor discharge and was discharged by it to the inlet of the molecular sieve driers, where it joined the gas from the first stage of the other 2 reciprocating compressors. The SEPCO compressor was used mainly as a back-up.

JT flash drum and other condensate collecting vessels

F.15 Condensate in the gas leaving the separators was knocked out in the condensate knockout drum and pumped back to the separators by 2 condensate transfer pumps. There was a high level trip on the drum, which would shut the ESVs at the inlet of the separators, to prevent carryover of condensate into the flare system. The condensate knockout drum was located north of the test separator at the east end of B Module and the condensate transfer pumps were next to the drum.

F.16 The condensate suction vessel, 2-C-202, collected condensate from the centrifugal compressor suction scrubbers. The level of condensate in the vessel was controlled by level control valve LCV 725. The condensate passed to the JT flash drum, entering the inlet pipe just downstream of the JT valve. The condensate suction vessel was located at the 68 ft level between the JT flash drum and the condensate injection pumps. The vessel was positioned as close to the ceiling of this level as possible in order to provide maximum net positive suction head to the condensate booster pumps. There was a balance line from the top of the condensate suction vessel to the header for the centrifugal compressor discharge scrubbers. The JT valve was in the ceiling of the 68 ft level at the extreme east side.

F.17 The JT flash drum, 3-C-701, located on the 68 ft level, was a condensate knockout and surge vessel which had somewhat different functions in the 2 modes of operation. In phase 1 operation it operated at a pressure of 665 psia and received condensate from the JT valve, PCV 721, and from the condensate suction vessel. The gas from the drum then passed to the suction of the second stage reciprocating compressors and the liquid to the condensate pumps. Somewhat more than half the condensate entering the drum came from the JT valve. A pressure differential of some 30 psi was maintained between the condensate suction vessel and the JT flash drum by differential pressure control valve DPCV 723A,B which controlled a flow of gas to flare. The pressure differential allowed condensate to flow from the condensate suction vessel to the JT flash drum. This vessel acted as a surge tank supplying the condensate pumps. The level of condensate in the drum was maintained by a level controller which controlled the speed of the condensate injection pump. There was a low level trip on the drum which stopped the condensate injection pumps to protect them against operation without any liquid intake. There was also a high level alarm, but no trip. This alarm was displayed on the local control panel and also as a common alarm in the Control Room.

F.18 In phase 2 operation the JT flash drum acted simply as a surge vessel for the condensate pumps and operated at a lower pressure, about 260 psia, and at 57°F. The JT valve was closed but served as a pressure relief valve to flare. DPCV 723A,B was set up in a different mode to act as a pressure control valve rather than as a differential pressure control valve.

Condensate injection pumps

F.19 Condensate from the JT flash drum was pumped into the MOL by a pair of condensate booster pumps in series with a pair of condensate injection pumps. The 2 condensate booster pumps, 3-G-701A,B, were centrifugal pumps; they raised the pressure to 670 psia and discharged to a common header. There was normally one pump operating and one on standby.

F.20 The condensate then entered the condensate injection pumps, 2-G-200A,B, shown in Fig J.9. The condensate injection pumps were single-acting, reciprocating, positive displacement pumps driven by an electric motor through a variable speed drive. They were supplied by Thyssen Maschinenbau Ruhrpumpen. The pump package consisted of an injection pump, an electric motor, a torque converter, a reduction gearbox, a control panel and a lubricating oil system; the torque converter was also referred to as the Voith coupling. The pumps had common suction and discharge headers.

F.21 Each condensate injection pump consisted of 3 horizontal cylinders, pistons, inlet and outlet valves, and suction and discharge manifolds. The reciprocating action alternately raised and lowered the pressure in each cylinder, causing it to fall below that in the suction header and draw in condensate through the suction valve and then raising it above the pressure in the discharge line into which it then flowed through the discharge valve; the 2 valves were spring-loaded to close.

F.22 The pump motor was a 500 hp constant speed induction motor with a nominal full load speed of 1725 rpm. The output shaft of this motor entered a torque converter. The output shaft from this converter passed into a gearbox which effected a 12.5:1 reduction in the rotational speed. The output shaft from this gearbox then drove the crank on the pump. The maximum pump speed was therefore 138 rpm. The torque converter, or Voith coupling, was a device by which power was transmitted from a driven input shaft to an output shaft by transfer of fluid between an impeller on the input shaft and a turbine wheel on the output shaft. The amount of torque transmitted was controlled by guide vanes which adjusted the flow of the fluid.

F.23 The pump speed, and hence pumping rate, was controlled by level controller LIC 720 on the JT flash drum. There was a selector switch on condensate control panel JCP 057 which allowed control to be exercised instead by the level controller of the condensate suction vessel, but it appeared to have fallen into disuse. The level controller altered the set point of the speed controller SC 501 on pump A or SC 502 on pump B; the speed controller then altered the guide vane setting of the torque converter. Panel JCP 057 was located just to the east of A pump.

F.24 Each pump was provided with an isolation or shutdown GOV on the inlet and another on the outlet, the suction and discharge valves on A pump being GOV 5005 and GOV 5006 and those on B pump being GOV 5007 and GOV 5008, respectively. On the suction side there was a manual isolation valve upstream of the GOV and a pulsation dampener downstream of it. On the discharge side there was a pulsation dampener, a high pressure trip and then an NRV upstream of the GOV.

F.25 The function of the GOVs was to effect automatic isolation of the pump. The valves were pneumatically operated ball valves, the suction valve being 8 inch and the discharge valve 6 inch. Each valve was an air-to-open valve which would close on loss of air pressure. The pumps had a number of trips which would cut off power to the motor. There were trips on low suction pressure, high discharge pressure, lube oil failure, seal failure, high motor winding temperature, high motor or pump bearing temperature and high vibration. A pump trip would also cause closure of the GOVs, thus isolating the pump. If a pump trip occurred so that the GOVs closed, it was necessary in order to restart the pump to reset the GOVs. This was done from panel JCP 057.

F.26 The function of the pulsation dampeners was to smooth out the pressure fluctuation caused by the reciprocating action of the pumps. They were essentially spherical vessels divided by a rubber diaphragm, which in normal operation was precharged on the upper side with nitrogen. Both suction and discharge dampeners had a volume of 75.7 litres.

F.27 There were 2 methods of electrical isolation of the pumps: locking off and racking out. Locking off involved locking off the isolation switch for the pump; the power from the 120 V AC UPS to panel JCP 057 remained on. Racking out involved pulling out the switchgear rather like opening the drawer of a filing cabinet; this cut off power to the panel. There was a manual pilot latch valve, or push-pull button, supplied by the power supply to JCP 057, which could be used to open the GOV. If there was power to the panel, the pilot latch valve when pulled would remain out, whilst if there was no power, it would not, and would need to be held out.

F.28 The pump local control panels JCP 043 and 044 were located at the north-east corner of each pump. The pump start buttons were at these panels. The suction and discharge GOVs on each pump were both on its south side, the discharge valve to the west of the suction valve, the 2 valves being about 2 ft apart. The push-pull buttons for the 2 valves were near the discharge GOV. Each pump had a lube oil package. Local alarms for the lube oil system on each pump were given on its local control panel.

F.29 There was no local alarm indication for low suction pressure or for high discharge pressure, either on panel JCP 057 or the local pump panels.

F.30 The discharge manifold was integral with the pump itself. The discharge line was taken off one side and the relief line off the other. In the original design the pressure safety valve was mounted on the pump itself, but as installed the PSV was on a relief line, as described below.

F.31 The pressure safety valves PSV 504 and 505, on pumps A and B respectively, were located in C Module. It was understood that the PSVs had been arranged in this way to prevent water reaching the valve and causing corrosion. There were manual isolation valves on the discharge lines from the PSVs. These lines then entered a common line, which had another manual isolation valve on it and which returned to the condensate suction vessel.

F.32 The relief line from the A pump to PSV 504 was line 2-P-524-4"-F15. The line coding indicates that the pipe was 4 inch diameter and pressure rating F15, which was the Bechtel code for a 900 lb rating. The corresponding working pressure was 2160 psi. The rating of the pipework flange on the upstream side of PSV 504 differed between drawings, being in some cases F15 and others G15, corresponding to 900 lb and 1500 lb rating, respectively.

F.33 Condensate from the discharge header of the condensate injection pumps passed through PCV 511. The purpose of this valve was to maintain a pressure sufficient to prevent flashing off of the condensate if the pressure in the MOL fell, essentially a pressure greater than that in the JT flash drum. PCV 511 was located towards the west end of the 68 ft level.

F.34 The volume of the pump system when shut in by the GOVs, taking account of the volume of the pulsation dampeners, comprised the volume of the pipe between the suction GOV and the pump, that of the pump itself, that between the pump and the discharge GOV (or strictly the NRV) and that of the relief line, and was some 400 litres. For a condensate density of 500 kg/m^3, the mass of condensate shut in would be 200 kg.

Methanol injection system

F.35 The main methanol injection pump, 3-G-702, was a 6-head injection pump and was located on the skid deck to the east of the drilling derrick and north of the deoxygenation towers. Plate 25(b) shows a photograph of the front view of the pump taken at Peterhead. For the installed pump this was the view looking north. The methanol supply came from a methanol tank, 2-C-201, which had a capacity of about 600 gallons and was kept filled from transportable containers. There was in addition an air-driven methanol pump, 2-G-201, the so-called "windy" pump, which supplied a further set of injection points. There were also 2 pneumatic pumps, the Williams pumps, which could be connected for use as back-ups. The location of the methanol injection points on the plant is shown in Fig J.8. On 6 July there were 2 injection points upstream of the JT valve, one the normal injection point fed from the main methanol pump head D and one a temporary injection point fed by a hose from head F.

Gas flaring

F.36 Gas between the production separators and the inlet of the centrifugal compressor system could be sent to flare through PCV 51/1,2. Gas from the outlet of that system could be sent to flare through PCV 1000A,B. There was at the same point a take-off of gas for fuel gas. Gas between the JT flash drum and the second stage reciprocating compressor system could be sent to flare through DPCV 723A,B. Gas from the second stage reciprocating compressors went 3 ways; to serve as lift gas, to MCP-01 and through PCV 945 to flare.

Control Room

Condensate injection pump displays and alarms

F.37 The status of the condensate injection pumps was indicated in the Control Room by lights on the mimic panel. There were 2 status lights, a red running light and a green shutdown light, and normally one or other of these lights would be on.

There was also on the mimic panel an amber alarm light which came on whenever there was a change in the status of a pump. For example, the amber light would come on if the pump was running and then stopped. In this case the green stop light would also come on. If the operator accepted the alarm, the buzzer would cease and the amber light would stop flashing. If he tried to reset the alarm, the amber light would still remain on because the alarm condition still existed. The amber light might remain on even if equipment was electrically isolated, because it had a separate electrical supply. The effect of stopping the pump and effecting electrical isolation by locking off a pump would be that the red running light would go out and the amber stop light would appear on the mimic panel. The effect of stopping the pump and effecting isolation by racking out would be to extinguish both red and green lights. The amber light would be illuminated when the pump stopped but would go out if the alarm were accepted and reset. An amber light on a pump could mean one of 3 things: that the pump had stopped; that it had been isolated by locking off; or that it had been isolated by racking out and that the alarm had not yet been reset.

Electrical supply system

Uninterrupted power supplies

F.38 Tables of the items supplied by the UPS systems were given in the Petrie Report (Tables 1-4). These had been reviewed by Occidental and were confirmed to be comprehensive. The items supplied by the D Module 125 V DC UPS are shown in Table F.1. They included emergency lighting in the accommodation and the HVAC system and dampers. They also included post lube oil pumps which supplied lubrication during the rundown period of certain items of rotating equipment. Table F.2 lists the items supplied by the D Module 120 V AC UPS. They included the general alarm/personal address (GA/PA) system and emergency telephones as well as the F&G system, which also included the solenoid valves for automatic activation of the fire water deluge sets. There were 2 further UPSs in the Utility Module, a 125 V DC and a 120 V AC UPS. The items supplied by these are listed in Table F.3.

Hazardous area classification

F.39 For the purposes of hazardous area classification the codes define 3 zones:

Zone 0 - A zone in which an explosive atmosphere is continuously present or present for long periods.

Zone 1 - A zone in which an explosive atmosphere is likely to occur in normal operation.

Zone 2 - A zone in which an explosive atmosphere is not likely to occur in normal operation and if it does will only exist for a short time.

A safe area is one in which an explosive atmosphere is not expected to occur. Hazardous area classification does not fully protect against ignition of a large leak, which may find an ignition source beyond the classified area.

F.40 Electrical equipment for use in Zones 1 or 2 is designed so that it does not constitute a source of ignition. The standard of safeguarding applicable to Zone 2 is lower than that applicable to Zone 1, because risk of a flammable mixture being present is less.

F.41 Diagrams showing the hazardous area classification of the platform were presented. On the production deck A-C Modules were Zone 2 areas, except for a small Zone 1 area in B Module near the production header, but for C Module the walkways at both the west and east ends and the air intakes of the centrifugal compressor turbines at the latter side were safe areas. D Module was a safe area. Most of the 68 ft level was a Zone 2 area, except for small safe areas at the north landing on the west side, the north-west corner, the north-east corner, and for part of the produced water area on the east side.

Gas detection system

F.42 The gas detection system is described in outline in Chapter 3. Further details are given here on the gas detection system in C Module with special reference to the information necessary to interpret the evidence on the gas alarms, including that bearing on the time delay of the detectors and the possibility of a detector not registering the gas cloud. Evidence on the gas detectors was given by 2 of the Occidental technicians responsible for the F&G detection system, Mr E Scothern and Mr D C Tea, and a representative of the gas detector manufacturers, Sieger Ltd, Dr D Balfour.

Location of gas detectors

F.43 The location of the gas detectors in C Module is shown in Fig J.10 and in Table F.4. On the height of gas detector G101/2 there was a conflict between the evidence of Mr Scothern and Mr Tea, the former putting it near the roof and the latter some 2-3 ft above floor level. Counsel to the Inquiry submitted that in so far as Mr Scothern had not been dealing with the system since 1987, whereas Mr Tea had, the latter's evidence was to be preferred.

Types of gas detector

F.44 The gas detectors were Sieger detectors types 770, 780 and 910. Type 910 was the most modern type and it was policy to replace any detector which fell to be replaced with this type. The features of the types 780 and 910 detectors were described by Dr Balfour. The principle of operation of the sensor was the catalytic oxidation of the hydrocarbon gas on a catalyst bead and the measurement of the change in resistance of the bead caused by the heat evolved in the reaction. The gas passed to the sensor through a sinter filter. The detector was held in a weather protection housing.

Composition and LEL of potential gas leaks

F.45 The streams which had potential to leak into the module were essentially natural gas and condensate. These are often approximated by methane and propane, respectively. The LELs of methane and propane are 5% and 2.1%. The actual compositions of the hydrocarbon streams at the suction of the first stage of the reciprocating compressors and of the second stage of the reciprocating compressors and at the discharge of the condensate injection pumps as given by Dr Balfour are shown in Table F.5. The LEL of a gas mixture may be estimated using the Le Chatelier equation. Dr Balfour's estimates using the Coward and Jones form of this equation were 3.54%, 3.81%, 2.16% and 2.34%, for streams at positions 170, 220 and 350, cases A and B, respectively. It may be noted that the LEL 2.16% for the stream at position 350 for case A is very close to the LEL of 2.1% for propane.

Gas detector settings

F.46 The gas detectors were calibrated for methane but were used to detect other hydrocarbons also. The low alarm setting was 15% of the LEL for methane and the high alarm setting 75% of the LEL. For methane these settings therefore correspond to concentrations of 0.75% and 3.75%, respectively. On a gas detector calibrated to read 100% full scale for 100% LEL of methane, the gas stream at position 350 would read 64.5% and 69.3% full-scale for cases A and B, respectively.

Gas detector dynamic response

F.47 There is a small time lag before a gas detector registers the gas concentration to which it is exposed. This lag is often characterised by the response time, the time for the reading to rise to 90% of its final value when subjected to a step change in the concentration. Dr Balfour gave the response times of the type 780 sensor as 19 and 24 seconds for methane and butane and those of the 910 sensor as 22 and 27 seconds for these 2 gases, respectively. An alternative parameter used to characterise the dynamic lag is the time constant. Taking from the above an estimate of the response

432

time for propane of 23 seconds, the corresponding time constant is some 10 seconds. The actual response of the detector to a gas cloud depends on the way in which the concentration changes with time. If what the detector sees is a sudden step change, it responds rapidly. For a change in concentration from zero to 15% LEL the times to low gas alarm given by Dr Balfour for both types of detector on both gases were less than 2 seconds. If the detector sees a ramp, or linearly increasing, input of concentration, then, after an initial transient, its output lags its input by a time equal to the time constant. The figures given by Dr Balfour apply to new detectors. Dr Balfour stated that detectors brought back from the field and tested again in laboratory conditions had behaved as did new detectors, but neither he nor the Occidental witnesses questioned were able to give any information on the dynamic response of detectors tested in the field.

F.48 Since the principle of operation of the detectors used was the measurement of the heat evolved consequent on the catalytic combustion of the hydrocarbon gas with air, there was a theoretical possibility that if the detector were flooded with pure gas, so that the concentration passed almost instantaneously from zero to 100%, the detector might not register an alarm. Dr Balfour stated that in fact there is a delay introduced by the diffusion of the gas through the filter and that the detectors do respond even when flooded. The effect of a jet of liquid condensate was also considered. In this case Dr Balfour believed that the detector would be protected by its weather protection housing.

Gas detector reliability and disabling

F.49 The reliability of the gas detectors was explored both with Dr Balfour and with Occidental personnel. Dr Balfour referred to the blocking of the filters by salt crystals, wind-borne particles, water or even fire-fighting foam, and to contamination of the catalyst by silicon and other chemicals. Silicon poisoning had been a problem, but steps had been taken and the problem much reduced. Any failure of the detectors would be unrevealed and it was therefore necessary to test them periodically. Some field data which his company had obtained showed a mean time to failure of about 10 years. Mr Tea had experienced deterioration of detectors in the turbine enclosures due to heat and in the accommodation due to silicone polish sprays. Usually when a detector was out of calibration, it was possible to make a small potentiometer adjustment. Outright failure was rare, but he could not put a figure on it. The interval between calibration tests was 4 months.

F.50 Mr Tea explained that it was possible to disable individual gas detector zones by "pinning out", which involved inserting a pin into the module for that zone at the back of the control panel. This was not itself logged, though the work being done in the zone would be.

Emergency shutdown system

F.51 Various terms were used to describe a complete ESD of the platform, including platform emergency shutdown (PESD) and overall emergency shutdown (OESD), the latter being used particularly in the phrase "electrical OESD". The 2 had essentially the same meaning and are referred to here as PESD.

Activation of PESD

F.52 Although there were separate pneumatic and electrical ESD systems, activation of one resulted in activation of the other so that the final effect was the same. Pneumatic PESD was initiated by loss of pressure in the pneumatic pressure loop due to melting of a fusible link. It was also activated by the action of the electrical ESD system. De-energisation of the latter caused depressurisation of the pneumatic pressure loop by activation of solenoid valves. A third way in which the pneumatic loop might be activated was loss of instrument air pressure. The electrical OESD, or PESD, system consisted of a bank of relays in the Control Room which were held energised when the plant was operating. They were de-energised by loss of power from the 125 V DC system. The electrical OESD was also activated by the action of the pneumatic ESD system by de-energisation of the relays. It was stated that loss of the main power supply would cause a PESD, but the mechanism by which this occurred was not clearly established.

F.53 PESD was activated automatically by a limited number of major process upsets. An example given was high pressure on the MOL, caused perhaps by closure of a valve at Flotta, which would trip the MOL pumps and lead to a PESD. On the other hand shutdown of a major item of equipment did not necessarily involve a PESD. High level in one separator would cause shutdown of that separator and of its associated wells, but not shutdown of the platform.

F.54 As far as concerns fire, there was no mechanism other than the fusible links by which fire would activate the PESD. Neither a gas alarm nor a fire alarm would in itself initiate an ESD.

F.55 Detection of gas at equipment located in a safe area activated shutdown of that equipment. This applied to the main generators and in this case the loss of main power would lead to a PESD.

F.56 PESD could be activated manually from the Control Room or from manual push-buttons (break-glass time switches) at 20 locations on the platform. The procedure was that anyone aware of a possible hazard should contact the Control Room, but the purpose of having manual ESD points distributed around the platform was so that personnel could effect shutdown without having to communicate with anyone else and all operating personnel had the authority to initiate a PESD.

Effects of PESD

F.57 One effect of an electrical PESD was to depressurise the pneumatic pressure loop and so initiate a pneumatic PESD also. Likewise, one effect of a pneumatic PESD was to de-energise the electrical PESD system.

F.58 On PESD the wells were shut down by closure on each well of the downhole safety valve (DHSV), the hydraulic master valve (HMV) and the wing valves; the first 2 closed on loss of hydraulic pressure and the latter on loss of pneumatic pressure. A PESD involved the shutdown of all major items of process equipment such as production separators, gas compressors and pumps and closure of all the process ESVs. A PESD caused closure of the ESV on the MOL but closure of the ESVs on the gas pipelines was by manual push-button.

Blowdown on PESD

F.59 Although PESD initiated blowdown of inventories from equipment by opening blowdown valves to the flare system, there were exceptions. Some major items such as the centrifugal compressors were designed so that on tripping they would isolate and blow down automatically. Other items such as the reciprocating compressors did not blow down on tripping, but did blow down on PESD. The production separators would blow down automatically only if the air pressure to the blow down valve on the separators was lost. The reason for not making this blowdown automatic on PESD was concern for carryover of liquids into the flare. The same applied to blowdown of the JT flash drum. The GCM blowdown had to be initiated manually, the reason being that this system contained a good deal of condensate and there was concern about dumping this to flare.

Other features of PESD

F.60 During the PESD the main generators remained on line but switched automatically to diesel firing on falling fuel gas pressure. Other systems which continued in operation were the instrument air compressors, the electrically driven utility and utility/fire pumps and other utility and safety systems.

F.61 On the other hand, the main generators were provided with gas detectors which would shut them down on detection of gas at the high alarm level. The shutdown of the main generators would de-energise the electrical system and thus result in a PESD.

F.62 The fact that the system was designed so that the instrument air compressors continued in operation in a PESD meant that there would not normally be a loss of air to those valves which were pneumatically operated and that the fail-safe action which would cause valves to open or close on loss of air pressure would not come into play.

ESVs on pipelines

F.63 Of the 4 pipeline ESVs, ESV 208 on the MOL was located in B Module and was an electrically operated MOV powered from the emergency switchboard. It had pneumatic back-up to close it on loss of electrical power and further back-up of nitrogen from an accumulator to close it on loss of air. These arrangements were a retrofit. Evidence on the retrofitting of this valve and that on the Claymore line was given by Mr A C B Todd, the maintenance superintendent. The valve was completed, tested and commissioned on 25 April 1988. There was outstanding the fitting of an "Add-on pack" to provide an interlock to shut down the MOL pumps if both valves were less than 75% open. However, the valve had not been formally handed over from construction by 6 July. Mr Wottge stated that the valve had operated satisfactorily in its shutdown mode when a faulty relay in the ESD system caused closure of all the pipeline valves. ESV 501, the ESV on the Claymore pipeline, was also an electrically operated valve with a pneumatic back-up to close it on electrical power failure and a further nitrogen back-up to close it on loss of air. This valve too had been retrofitted in early 1988. Mr Todd said that it was completed, tested and commissioned on 9 April 1988, but had not been formally handed over by 6 July. In early July a new ball valve was fitted to ESV 501. The Tartan pipeline ESV, ESV 6, was a hydraulically actuated valve with nitrogen back-up. The MCP-01 pipeline ESV, ESV 956, was also a hydraulically actuated valve with nitrogen back-up.

Pipeline depressurisation facilities

F.64 There were on the 4 platforms facilities for depressurising the 3 gas pipelines by flaring the inventories, but they were limited by the gas flows which could safely be flared and such depressurisation normally took days rather than hours. All 3 pipelines could be depressurised at the Piper end by making the necessary connections and opening hand valve HCV 961 (see Fig 3.10). This valve was located near the pig traps. It was understood that about 100 MMSCFD could be passed through this valve. The Piper-MCP-01 line could be depressurised at MCP-01 by opening pressure control valves PCV 4353A,B to the blowdown skid. The depressurisation of the Piper-Claymore line could be effected at Claymore by opening hand control valve FCV 970. The Tartan-Piper line could be depressurised at Tartan through a valve. The normal rate of depressurisation was said to be about 12 MMSCFD with a maximum of 30 MMSCFD.

Phase 1 operation and GCM changeout

F.65 Preparations for the GCM changeout were made by Mr A Carter assisted by Mr T A Henderson, a lead production operator. Between 28 June and 5 July the 2 were on the platform together. Mr Henderson left on 5 July, but Mr Carter stayed on to oversee the changeover. No comprehensive work pack for the changeover was recovered, but Mr Henderson assembled a number of documents which he said Mr Carter had prepared. The latter had produced documentation covering the changeover from phase 2 to phase 1, operation in phase 1 mode, work to be done during the changeover period, advice to operators on this work and restoration of phase 2 operation. The work pack included lists of valves to be closed and of spades to be inserted. The pipes into the molecular sieve driers in the GCM were to be spaded off, since men would be working in the driers, but the GCM itself was not to be spaded off. The work pack also included instructions on depressurisation of equipment and on methanol injection. There were also several control loops which needed to be adjusted for operation in phase 1 mode. The setpoint on the JT valve, PCV 721, had to be changed so that it would control at the different pressure. The transmitter on DPCV 723A,B required to be switched so that the loop would operate to control differential rather than absolute pressure.

F.66 The GCM was taken out of service on Sunday 3 July. The gas plant was shut down and the compressors depressurised. The equipment and pipework in the GCM were then depressurised with the exception of the line to the SEPCO compressor; valves 30 and 62 on this line were closed. The teams carrying out the isolations were led by Mr Carter and Mr Henderson. The work programme for the GCM was scheduled for the period 3-15 July. One major item was the changeout of the beds of the molecular sieve driers. Since the beds adsorbed hydrogen sulphide as well as water, this was an operation liable to give rise to gas smells. There were various planned maintenance jobs and work on orbit valves.

Status of certain structural features

F.67 The status of certain structural features on the platform in early July is relevant in that it bears on the possibilities for the spread of flammable gas and of fire.

F.68 One such feature was the possible existence of apertures in the firewall between B and C Modules. It was alleged that part of the firewall near the door had been removed in order to allow work to be done on pipes passing through the wall. Several passages of evidence were heard on the point. It was agreed that a hole had been made in the firewall to allow painters to do needle-gunning work. However, whereas it was originally stated that there was a hole 5m x 4m in the firewall above the door towards the west end of the module, the final outcome was that the wall had been largely restored, although by 6 July an annular gap of perhaps 1-2 inch remained around at least one of the pipes penetrating the wall and over an area of uncertain size the fireproofing had not been remade.

F.69 There was also a door in the firewall opposite the MOL pig trap (see Fig J.3(c)); the door had a self-closing mechanism in the form of a weight on a chain enclosed in a tube. Evidence on this reduced to the allegation that on one occasion it was difficult to shut. There was a proposal to put a new access door in the wall to give access to the middle metering stream to allow removal of the turbine meter, but this work had not started by 6 July.

F.70 Evidence that the prover loop had been completely removed and other evidence that there was some scaffolding at the 68 ft level more or less below the area of the prover loop led to exploration of the possibility that some of the deck plates on the floor of B Module may have been missing. However, removal of the prover loop would not in itself create a gap in the deck plates and no evidence was given that such a gap existed on 6 July.

F.71 Some of the drawings of C Module (eg Fig 9.21 of the Petrie Report) showed a partitioned area at the west end of the module. The evidence was that there was no such partition at the main 84 ft level in the module.

Table F.1 - Items supplied by 125 V DC UPS in D Module mezzanine level

1. Emergency lighting for GCM, Utility Module, distribution boards EL 1, 2, 3 and 4, AAE, ERQ and LQW.
2. Turbine generator panels.
3. SPEEM + AAW distribution board.
4. Centrifugal compressors lube oil system.
5. High voltage and low voltage switchgear.
6. Main process control panel and MOL control panel.
7. HVAC panel and dampers.
8. Fire protection units P102A,B.

Note:
Based on Table 1 of Petrie Report (following para 4.3.6.1).

Table F.2 - Items supplied by 120 V AC UPS in D Module mezzanine level

1. General alarm and personal address system.
2. Main fire and gas panel.
3. Emergency telephones.
4. UPS shutdown contactor panels.
5. Divers' communication system.
6. Main control panel, MOL and gas separation panels.
7. Turbine and generator panels.
8. Drilling module fire alarm panel.
9. SPEEM PESD panel.
10. Turbine gas detection (1-P-102A,B).
11. Discharge scrubbers D/P valve (1-K-105C).
12. Condensate control panel JCP 057.
13. Metering and pig launcher and receiver local panels.

Note:
Based on Table 3 of Petrie Report (following para 4.3.6.1).

Table F.3 - Items supplied by 125 V DC UPS and 120 V AC UPS in Utility Module

A. 125 V DC UPS
1. 13.8 kV closing, tripping and indication supplies for 4-P-801 switchboards.
2. 4.16 kV closing, tripping and indication supplies for 4-P-802 switchboards.
3. 440 V tripping and indication supplies for 4-P-803 switchboards.

B. 120 V AC UPS
1. Fire and gas panel.
2. GCM local control panels.
3. Solartron telemetry system.
4. General alarm system.
5. Reciprocating compressors control panel.
6. HVAC control panel.
7. Flare control panel.
8. Depressurisation valves and solenoid valves.

Note:
Based on Tables 2 and 4 of Petrie Report (following para 4.3.6.1).

Table F.4 - Gas detectors in C Module

Area	Detector	Height	Location
C1	G22	3 ft down[a]	
	G23	3 ft down	
	G24	3 ft down	
	G25	3 ft down	
	G100/1	6 ft up[b]	
	G100/2	7-8 ft up	
C2	G101/1	12-13 ft up	
	G101/2	2-3 ft up	
	G101/3	15 ft up	
C3	G26		At ventilation fan inlet
	G27		Outside turbine compartment[c]
	G28		In compressor compartment
	G102/1	Roof level	At turbine intake
	G103/1	Below grating	At fuel gas valve
	G102/2[d]	Roof level	At turbine intake
C4	G29		At ventilation fan inlet
	G30		Outside turbine compartment[c]
	G31		In compressor compartment
	G102/2[e]	Roof level	At turbine intake
	G103/2	Below grating	At fuel gas valve
	G102/3[f]	Roof level	At turbine intake
C5	G32		At ventilation fan inlet
	G33		Outside turbine compartment[c]
	G34		In compressor compartment
	G102/3[g]	Roof level	At turbine intake
	G103/3	Below grating	At fuel gas valve
	G102/4	Roof level	At turbine intake

Notes:

(a) Down from ceiling.

(b) Up from floor.

(c) Detector outside compartment but with sample tube into compartment.

(d) Detector shared with area C4.

(e) Detector shared with area C3.

(f) Detector shared with area C5.

(g) Detector shared with area C4.

Table F.5 - Calculated composition of gas from selected potential leak points in C Module

Position	Composition (% v/v)			
	170	220	350 (case A)	350 (case B)
Gas				
Methane	65.7	71.6	20.0	21.4
Ethane	17.1	16.0	18.9	20.1
Propane	12.4	9.9	31.3	33.4
Butane	3.5	2.1	17.3	18.5
Pentane	1.1	0.5	10.0	6.1
Fraction 125-127	0.2	0.0	2.4	0.5
Fraction 175-365	0.0	0.0	0.1	0.0

Notes:

(a) Positions 170, 220 and 350 are at the first stage reciprocating compressors suction, the second stage reciprocating compressors suction and the condensate injection pumps discharge, respectively.

(b) Case A is for stream completely vaporised and case B for stream partially vaporised.

Appendix G

Supplementary Material on Chapters 5 and 6

Firewall failure

G.1 The analysis by Dr Cox of the over-pressures required to destroy the firewalls in C Module was described in Chapter 5. Further details of this analysis are given here.

G.2 The B/C firewall was a single-layer 4.5 hour integrity wall. The wall extended along the length of C Module from east to west and vertically from the production deck to the truss upper beam at a height of 6.35m. It consisted of an array of rectangular panels of 9.5 mm thick Durasteel 3DF2. The panels were of 2 main sizes and were each bolted into a rectangular frame which was a welded fabrication of 50 mm x 50 mm angle-section steelwork. Adjacent frames were bolted together, forming a "lattice". The lattice was typically 3 frames high. The lower edge of the bottom frame was continuously fillet-welded to the production deck and the upper edge of the top frame was attached to the underside of the upper truss beam by an arrangement of bolted and welded joints. The wall was further supported by clamping to the truss columns, the clamps being simple straps bearing on cleats which were site-welded to the lattice. The firewall is illustrated in Fig 5.2; the figure is schematic and is not to a consistent scale. The panelling is on the near side of the lattice. The figure shows 2 bays of the firewall with 3 vertical and 2 inclined members, all part of the truss, with lattice work and with panels, 3 high, bolted to it. The view in the figure is that seen from the inside of C Module looking south.

G.3 Information on the strength of the Durasteel panels and of the panel bolts was sparse and it was necessary to make assumptions. Durasteel 3DF2 is a composite material consisting of 0.5 mm perforated steel skins around a fibre reinforced cement core, with a total panel thickness of 9.5 mm. It was treated in the analysis as a homogeneous material with the same bulk properties as the composite sheet. A physical test was carried out at Aberdeen University and numerical modelling of this test gave reasonable agreement. Throughout the firewall 3/8 inch Whitworth bolts were specified but the steel grade was not known. The ultimate tensile strength of the grade assumed as representative of mild steel bolts was 432 MPa and a failure strength of 260 MPa was assumed throughout, this being representative of mild steel bolts. Further consideration led to a revision of the bolt strength. The assumed tensile and shear strengths were revised to allow for the thread form. The revised capacity of the bolts was calculated as 11.7 kN under tensile load and 6.7 kN under shear load. The maximum spacing allowed between panel bolts was 15 inches and between frame bolts 24 inches. The number of bolts was calculated from these figures. It would not be usual for there to be drawings and so the bolt spacings were subject to some uncertainty. The strength of the clamps was taken as 23 kN per clamp.

G.4 The C/D firewall was a triple-layer 6 hour integrity wall. The wall extended along the length of C Module from east to west and vertically from the production deck to the truss upper beam at a height of 6.38m. This wall differed from the single-layer wall in that the panels consisted of 3 identical Durasteel 3DF2 plates each 9.5 mm thickness and separated by 45 mm thick of dense mineral wool; the frames were smaller, being 7 rather than 3 high; there was a complex offset bolting arrangement; and the arrangement of the panel and frame bolts was different in detail. The firewall was clamped to truss 6 only by light duty hook clamps quite different from the clamps used on the single-layer wall. The triple-layer firewall is illustrated in Fig 5.3; the figure is again schematic but in this case the panelling is on the remote side of the lattice. The view in the figure is that seen from the inside of D Module looking south.

G.5 In the analysis of failure of the single-layer firewall the following failure modes were considered: panels, panel bolts, lattice framework, frame bolts, clamps and welds to the deck and to the truss. Failure of the panels was studied using both static and dynamic finite element techniques. Depending on the assumptions made, failures of a large panel (2.34m x 1.42m) under static loading were found to occur at 0.10-0.22 barg and under dynamic loading at 0.15-0.36 barg. The panel bolts were found to fail in shear loading at a pressure of about 0.1 barg and the frame bolts in tensile loading at a similar pressure. The clamps would fail at a pressure of about 0.12 barg. The lattice would collapse by formation of plastic hinges at about 0.53 barg. Failure of the welds was calculated to occur only when the pressure on the firewall was 4.7 barg. A similar analysis was made on the triple-layer firewall.

Wind tunnel tests

Gas detectors and gases tested

G.6 A brief account of the gas detection system in C Module is given in Chapter 3 and a fuller description in Appendix F. The location of the gas detectors in this module is shown in Fig J.10 and Table F.4. The description of the gas detection system in Appendix F covers the types of detector used; the gas mixtures which might occur as a result of leaks; the LELs of such gas mixtures; the settings of the detectors; their dynamic response; their reliability; and disabling of detectors. Attention is drawn to 3 points discussed more fully in that Appendix: the conflict of evidence on the height of gas detector G101/2; the time lag in the response of the detectors; and the practice of disabling gas detector zones by pinning out. No evidence was heard that any zone was pinned out on 6 July.

G.7 The gases the dispersion of which was simulated in the wind tunnel tests were propane and a neutrally buoyant mixture of methane, ethane and propane. The gas detector setting data used in the wind tunnel test experiments were as follows:

Gas	LEL	Concentrations (%)	
		Gas detector settings	
		Low alarm	High alarm
Methane	5.0	0.75	3.75
Propane	2.1	0.5	2.5

Background to tests and preliminary tests

G.8 The wind tunnel tests were performed by BMT Fluid Mechanics Ltd. at their wind tunnel at Teddington. A wind tunnel is used to perform small scale experiments on fluid flow. The object of interest is placed in the wind tunnel, the flow of air through the tunnel is set in accordance with principles of scaling, and the flow patterns are observed. It is a powerful and versatile device for studying flow of fluids around objects of complex geometry. Two wind tunnels were used, the main Environmental Wind Tunnel, and a smaller wind tunnel. The tests were performed on the 1: 100 and 1: 33 scale models used in the Inquiry, Models A and B, respectively, the models being taken away to the wind tunnel facility for the purpose.

G.9 The main series of tests were conducted on the 1:33 scale model, but as a preliminary to these tests, it was necessary to establish the ventilation air flow corresponding to the conditions at Piper on the evening of 6 July. This was done as follows. First, the flow-pressure drop characteristics of the 1:33 scale model were determined. The 1:100 scale model was then modified. On the original model the modules at the 84 ft level were represented by solid walls. For B and C Modules these were replaced by models of the modules similar to but simpler than the modules in the 1:33 scale model. The flow resistance of the C Module model in the 1:100 scale model was then adjusted to correspond to that measured on the 1:33 scale model. The 1:100 scale model so modified was placed in the larger wind tunnel and the air flow was adjusted to simulate the conditions at Piper. The wind conditions were based on those recorded by the *Lowland Cavalier* (see paras 3.138; also 3.3) and were taken as wind direction 207° and wind velocity 8.2m/s and for these conditions the ventilation

442

rate through the module was 46 m³/s. This corresponds to an air change rate of 39 air changes/h and to average air velocity of 0.5m/s. Other wind conditions were also studied and ventilation rates obtained as shown in Table G.1. With the ventilation rate thus established, the main tests were then performed using the 1:33 scale model in the smaller wind tunnel. A video of the tests on both models in the 2 wind tunnels was shown to the Inquiry. The flow through the model represented a speeded-up version of the flow in the actual module, 10 seconds on the model corresponding to 50-60 seconds at full scale. Propane was simulated in the tests using a mixture of argon and Halocarbon 12 and neutrally buoyant gas using a mixture of helium and carbon dioxide. Concentration measurements of these gases were taken at suitable sample points in the model using fine thermal conductivity aspirating probes. The number of sample points used varied between 5 and 27 per series.

G.10 Two sets of experiments were carried out. The first set investigated a number of different leaks, with emphasis on leaks from the area of PSV 504. The second set was concerned with leaks of neutrally buoyant gas. The tests conducted and the results of the first and second set of tests are given in Tables G:2-G.4, respectively. For each set of leak conditions a series of runs was performed, but the number of runs varied. For some conditions it was desired to take samples at 20 or more points, but in order to avoid excessive disturbance to the flow pattern the number of probes was limited to 5 in a given run. Thus it was often necessary to perform 4 or 5 runs to obtain the coverage of sample points required. The results for each condition were therefore referred to as a series.

Limitations of, and uncertainties in, tests

G.11 There are several sources of potential inaccuracy in wind tunnel tests. The most fundamental is the scaling process itself. Other sources include possible deficiencies in the models tested or in the meteorological conditions specified for the test and inaccuracies in measurement. In wind tunnel testing the system of interest is studied using a scale model. The scaling process involved in extrapolating the results to full scale involves some inaccuracy. However, there is wide experience with wind tunnel tests conducted on this basis. Making a very rough estimate of possible errors in average concentration, time and mass of fuel, Dr Davies indicated that they might be some plus or minus 20%. The 1:33 model was not an exact model of the equipment in C Module. For example, the compressors were modelled as "boxes" whereas in reality they were complex items of machinery with pipework, valves, etc. Dr Davies did not believe this was a significant source of error; it might alter a time interval from 20 to 25 seconds.

G.12 In the experiments measures were taken using aspirator probes. The response time of these probes was about half a second to a second, in full scale time units. Some typical traces of the concentrations measured in the experiments are shown in Fig G.1. Several features are noteworthy. Firstly, there was an appreciable difference in the final steady-state values. For example, for the 2 runs in series 25 for sensor G103/1 the steady-state values are approximately 2.6 and 3.1, while for sensor G101/2 they are 1.8 and 3.2. The lateral spread of the cloud, and hence the readings of sensor G101/2 tended to show a greater variability. Secondly, there was a high level of noise on the final steady-state value, so that an alarm might be triggered even though the smoothed steady-state value was below the alarm limit. This occurred in series 28. Thirdly, the initial part of the curves constitutes effectively a ramp, rather than a step, forcing function. On the full scale the concentration measured by the gas detector would, after a short initial transient, tend to lag the actual concentration by a time equal to the time constant of the detectors, estimated as 10 seconds. This would apply particularly to the low level alarms. The lag would be rather greater where the high level alarm limit was close to the final steady-state value.

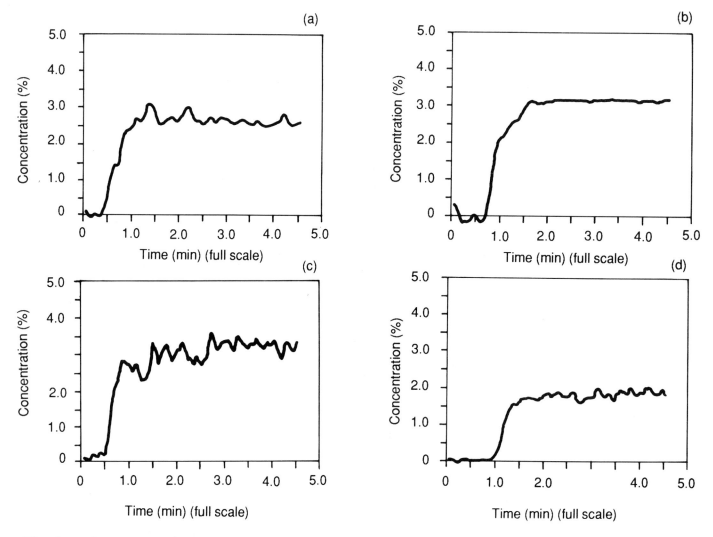

Fig. G.1 Gas concentration-time traces for sensors in repeated runs in series 25 in the BMT wind tunnel tests: (a) and (b) sensor G103/1; and (c) and (d) sensor G101/2.

G.13 It was possible that one or more of the gas detectors in C Module might not have been operational. In particular, the apparent failure of the 2 C2 detectors G101/1 and G101/2 to activate first constrained the interpretation of the test results. The possibility was explored that G101/1 might have been pinned out while work was done on PSV 504. However, there was no evidence that any detector was in fact pinned out.

Explosion simulations

G.14 Explosion simulations using the FLACS computer code were commissioned by the DEn and by Technica and a further run was commissioned by the Inquiry. The results of this work were described in Chapters 5 and 6. Further details are given here of the FLACS code itself, of the explosion simulations, and of the limitations of, and uncertainties in, these simulations.

The FLACS code

G.15 The FLACS code is designed to solve the fundamental equations of fluid flow taking into account turbulence and combustion. The 3-dimensional Navier-Stokes equations, suitably amplified to include the effects of turbulence and combustion reactions, are cast in discrete form, employing a finite volume technique, and are solved implicitly. Turbulence is modelled in terms of eddy viscosity and combustion in terms of turbulent, mixing-limited reaction. The space modelled is divided into a grid of "boxes" of volume one cubic metre. Normal assumptions are that the flammable gas cloud is a quiescent homogeneous stoichiometric mixture so that the effects of any concentration differences within the gas cloud, of any ventilation air flow or of a continuing leak source are neglected. Ignition is modelled as a weak ignition by assuming that at time zero half of the flammable mixture in one of the boxes has undergone combustion so that the temperature of the gas in the box is correspondingly increased. Details of the structure of the module and of the equipment contained in it are captured by a front-end code CASD. A further program is then used to process this information into a form in which it can be utilised by the FLACS code. The principal output of the code of interest in the present context is the explosion pressure generated, but the code also produces profiles of the concentrations of the unburnt fuel and the combustion products and of the gas velocities.

G.16 The effect of obstacles in the module is to enhance turbulence and this may have a strong influence on the pressures generated. Another important influence on these pressures is that of venting. Venting at open or partially open ends of a module is automatically taken into account in the code, but it is also necessary to allow for venting by wall failure. This is handled in the code as follows. The pressure at which a wall will fail is determined. It is assumed that the wall starts to move when this pressure is reached. The movement of the wall is then calculated from the mass of the wall and the pressure on it. It is further assumed that the distance which the wall travels will be limited by obstructions in the adjacent module. The movement of the wall opens up a gap between the wall and the floor and ceiling of the module and this vent area is expressed as an effective wall porosity. There is therefore an interaction between the pressure generated and the venting due to wall failure; the pressure causes wall failure and the wall failure acts to reduce the pressure.

G.17 The FLACS code has been validated by comparison of results obtained from the code with measurements made in experiments on explosions in scale models of modules. The models used were on scales of 1:33 and of 1:5. A typical explosion experiment is shown in Plate 26(a). The over-pressures predicted by the code lie within plus or minus 30% to a confidence level of some 95% of those measured experimentally. The variability of model experiments themselves is of the same order. It was Dr Bakke's expectation that the measured over-pressures would tend to be greater in full scale tests. The work at CMI has been sponsored by a number of organisations and a number of studies have been conducted on gas explosions in modules of offshore platforms.

Simulations performed for the DEn

G.18 Soon after the disaster the DEn commissioned CMI to carry out a series of simulations of explosion of idealised flammable gas clouds in C Module. Their report was issued as Annex 3 to the Petrie Final Report. The report, though dated October 1988, was based on the information available in August. CMI was provided with information on the geometry of C Module in the form of drawings and photographs. The plan view of C Module produced is shown in Fig 5.1. This figure also shows the pressure recording points, flammable gas clouds and ignition points. The Control Room was located in the mezzanine level of D Module and thus on the upper part of the C/D firewall at a point corresponding approximately to recording point 5. The firewall failure pressures were specified as 0.138 bar for the B/C firewall and 0.25 bar for the C/D firewall. The mass of the walls was given as 63 kg/m^2. It was agreed to assume a wall porosity of 20% for the B/C firewall and a 40% porosity for the C/D firewall. (In the original report and evidence it was stated that the porosity of both firewalls in this work was 20%, but this was corrected in the later report and evidence.) The compositions of the 2 fuels used in the simulations, natural gas and condensate vapour, are given in Table G.5, Sec A. For simplicity, the compositions used in the code are the equivalent mixtures given in Sec B of the table.

445

G.19 Five simulations were performed in this work. The simulations specified (cases 1-5) and the results obtained are shown in Table 6.2. Pressures were recorded at 8 points as shown in Fig 5.1. Points 1-4 are along the south wall and points 5-8 along the north wall. The points are at a height just over halfway up the module. For case 1 the pressures generated were sufficient to cause failure of both firewalls at all the recording points P1-P8. The pressures in case 2 were appreciably higher, those in case 3 higher than in case 2, and those in case 5 higher still, so that all these cases would cause failure of both firewalls at all the recording points. For case 4 the pressures were on the borderline of those required to cause failure, exceeding the firewall failure pressure only at points P2-P4 on the south wall. In particular, the pressure point 5 near the Control Room was below the failure pressure. Although this work was in large part superseded by that commissioned by the Inquiry, the apparent trends which it illustrates are important. One is that the pressures generated by a cloud of condensate as opposed to natural gas are somewhat higher (case 3 v case 2). Another is that an ignition source located at the centre of the module gives appreciably higher pressures than one located at the end (case 2 v case 1). A third is that the pressures occurring in the absence of venting by firewall failure are much higher (case 5 v case 3). And finally there is the not unexpected result that a smaller gas cloud gives rise to lower pressures (case 4 v case 1).

Limitations of, and uncertainties in, simulations

G.20 The simulation of an explosion in a module is a complex undertaking and the technology has been developed only recently. The model used in the simulation involves a number of idealisations and assumptions. There are some potential sources of inaccuracy in the model itself and in the solution of the model equations. Questions also arose concerning the input data for the particular scenario modelled for the Piper explosion. The idealisations made were those normally used in the code and have already been described. The flammable gas cloud had an idealised rectangular geometry and was assumed to be homogeneous. No account was taken of air flow through the module, of the effect of a continuing gas leak or of any upwind movement of the cloud. Some work has been done by CMI on gas cloud homogeneity. A high-momentum, quite large release tends to fill the module with a cloud which is relatively homogeneous and which rises in concentration as the release proceeds. With regard to the effect of ventilation, the air velocity in C Module was of the order of 0.5m/s, some 2-3 orders of magnitude below the highest velocities occurring in the explosion. Some inaccuracy is introduced during the process of integrating the differential equations of the model. The model used includes some parameters that can be "tuned" to fit experimental results obtained at small or medium scale. In such a case, however, there must always be some uncertainty when the model is extrapolated to full scale. Another idealisation was the modelling of the process of failure of the firewalls, which assumes that once the failure pressure is reached failure is effectively total and instantaneous. In fact processes such as shearing of the bolts must take some finite time. There remained some uncertainty concerning the extent to which the model took into account phenomena such as an external explosion. These were described by Dr Chamberlain in Part 2 of the Inquiry but were not explored with Dr Bakke in Part 1. To some extent the tuning process mentioned may allow for such phenomena, but since they tend to be more important at full scale, the allowance may not be fully adequate.

G.21 Questions concerning the input data for the Piper explosion centred particularly around the location of the ignition source, the porosity of the firewalls and the behaviour of the ducting around the centrifugal compressors. The selection of the location of the ignition source was not based on any information but was made to give something close to a worst case, ie one which would generate the highest pressures. By selecting this worst case it was possible to explore the smallest size of flammable cloud to give an explosion. The effect of the porosity of the firewall was somewhat reduced by the fact that according to the model only a small fraction of the ultimate porosity is developed at the time when the pressure reaches its peak. Thus although the treatment of porosity was very approximate, provided some reasonable allowance is made, the effect on the final results may not be great. The available vent area may also be increased by the destructive effect of the high wind velocities generated in an explosion. In particular, the ducting of the centrifugal compressors at the east end of C Module would be vulnerable to such winds. Loss of this ducting was not allowed for in the simulation.

Rating of upstream flange on PSV 504

G.22 As described in Chapter 6, there was uncertainty about the rating of the upstream flange on PSV 504. A summary of the evidence on this is given here.

G.23 A large number of drawings and other documents which bore on the question were produced by Mr M Skidmore, a Senior Facilities Engineer with Occidental. Many of these were documents related to the original design and pre-dated the construction of the condensate injection system. They showed that the original specification of the upstream flange was 900 lb. However, there were a pair of "as-built" drawings which for PSV 505 gave both on the drawing and the material code a 1500 lb rating and for PSV 504 a 1500 lb rating on the drawing but a 900 lb rating in the materials list. Mr Skidmore also produced documents relating to later modifications to features such as the pulsation dampeners which gave a 900 lb rating. There was in addition a telex dated 22 August 1977, and thus in the construction phase of the condensate system, calling for urgent action on shortage of materials, including 900 lb flanges. It was put to Mr Skidmore that in this situation a 1500 lb flange might have been fitted to progress the work, but he discounted this as poor practice.

G.24 Another source of information on the flange rating was the safety relief valve (SRV) test certificates for the periodic recertification work. A collection of certificates for PSVs 504 and 505 was examined. A summary of these certificates is given in Table G.6. Valve technicians who had worked on these PSVs since 1985 were called, namely Mr J Tait, a service engineer with Score (UK) Ltd, Mr A Pirie, a service engineer with Wood Group Valves and Engineering Services Ltd, Mr R W Barclay, formerly a valve technician with the same company, and Mr A C Bruce, formerly a valve technician with Score. Mr Tait described the sources from which a flange rating for an SRV, or PSV, might be obtained. There were history cards, test certificates, and site inspection. The flange rating might be obtained from the nameplate, if that was still legible. Alternatively, the flange could be measured and the rating looked up in flange tables. There was also an Occidental printout of the valves, but while this gave the pipe diameter, it did not necessarily give the flange rating.

G.25 As Table G.6 shows, the first entry of the upstream flange on PSV 504 as 1500 lb was on 29 October 1985. This was a rush job which Mr Pirie did alone, making the adjustment on the scaffolding using test flanges, and filling in and signing the test certificate No 3825. The job was too short to fit blind flanges. He did a similar job on PSV 505, recorded on test certificate No 3826. In both cases he recorded the upstream flange rating as 1500 lb. He could not remember how he established this; he did not think he had the previous test report available. He must have been sure of the rating or he would not have put it down, but he could have been mistaken. The next overhauls of PSV 504 and 505 were done by Mr Barclay and Mr J McDonald. The test certificate No 0101 for PSV 504 dated 16 September 1986, filled in and signed by Mr Barclay, again showed a 1500 lb rating. Mr Barclay said he would not have completed any part of the certificate before doing the job. He could not recollect how the flange rating was determined, but it would have been necessary to know it to do the job; in fact in this particular case it was necessary to improvise a test flange. When he came to sign the certificate he must have thought the rating entered was correct. The next day PSV 505 was overhauled, the overhaul being done by Mr McDonald and the test certificate No 0104 being filled in by Mr Barclay and signed by Mr McDonald; again an upstream flange rating of 1500 lb was entered. On 24 March 1988 PSV 505 was overhauled by Messrs Tait, Bruce and Sutton. The certificate

No 2607 is signed by Mr Bruce, though both he and Mr Tait said that it was filled in by the latter. Mr Bruce said that he and Mr Sutton measured the upstream flange rating on both PSVs and found it to be 1500 lb. They also checked against the certificates for the previous tests, that for PSV 505 being No 0104 on 17 September 1986. They had taken out both certificates, since they did not know which of the 2 PSVs was to be worked on and there was no difference in the flange ratings, both were 1500 lb. He believed that to get ahead with the paperwork the box on the certificate containing the entry on the flange rating was filled in before the work was done. Mr Tait, who filled in the certificate, was unsure where he obtained the flange rating. It was his practice to read the previous test reports and write notes in a notebook. He could have transferred information from his notebook to the new certificate, though if any inconsistency had been noticed he would have gone and measured the flange. He also stated that there could have been several valves being done that day and that since this overhaul itself was completed only about 22.00 hours, the certificate was probably signed the next day. It was put to Mr Skidmore that these test certificates showed that the flange rating was 1500 lb, but he maintained his belief that the flange had been originally specified as 900 lb and had not been changed. He agreed that if the 1500 lb flange rating on these certificates was a mistake, it was a separate one from that on the as-built drawings.

G.26 Photographs of PSV 504 and 505 were available to the Inquiry. The photographs were provided by Mr T Rogers, Facilities Engineer with Occidental, and were believed to have been taken about 1978. An attempt was made to determine the flange sizes from these photographs by Mr D J Turner, Managing Director of Camera Alive Ltd, using computer matching. Mr Turner stated that he obtained a good fit assuming the downstream flange to be 600 lb and the upstream one to be 900 lb, but a poor fit assuming the latter to be 1500 lb. Counsel to the Inquiry indicated that he did not accept this evidence, but did not pursue the matter by cross-examination.

G.27 The evidence on the rating of the upstream flange on PSV 504 was therefore contradictory. Fitters who had done the most recent work on the flange had entered it as 1500 lb on the test certificates. Although in principle this evidence might be preferred to that of the original design documents showing it to be 900 lb, there was doubt whether the entries were based on the immediate knowledge of those working on the flanges rather than transferred from previous test certificates. It is conceivable that the 1500 lb rating was entered in error during the work in October 1985 and then perpetuated. On the other hand the shortage of 900 lb flanges during construction and the rating shown in the "as-built" drawing are possible pointers to a 1500 lb rating.

Autoignition in relief line on A pump

G.28 In their third report Drs Richardson and Saville explored the possibility of rupture by autoignition in the relief line on A pump. The report included a detailed study of the possible effects of the estimated maximum explosion pressure of some 300 bara, which is described here. The effects studied were those on the pipe and, for a blind flange, on the flange itself and on the bolts and the ring.

G.29 The authors drew attention to the fact that if the time taken to reach the final pressure was substantially less than the response time of the container, the walls of the container would experience a transient stress double that experienced in a slow application of pressure, but stated that the response time of a rigid metallic container would be much less than the duration of the explosion and so no enhancement would take place. However, such enhancement would apply to bolts which were slack; in this case the maximum stress would be double that for tight bolts.

G.30 The relief pipe was 4 inch nominal bore. Since there was doubt about the pipe ratings, both 900 lb and 1500 lb ratings were considered. The Occidental specification for both ratings was for material to standard ASTM-A106, which gives a yield strength of 35000 psi and a minimum tensile strength of 60000 psi. The actual wall thickness was required to be at least 0.875 of the nominal thickness. When determining schedules Occidental made a corrosion allowance of 0.125 inch. Thus there were also 2 cases to consider with respect to pipe wall thickness. One was the nominal thickness and the other 0.875 x nominal thickness less the 0.125 inch corrosion allowance. The piping specifications considered and some results of the work are given in Table G.7. The specification C1D corresponds to Bechtel F15 rating and 900 lb flanges and C1E to G15 rating and 1500 lb flanges. Sections A and B of Table G.7 give, respectively, the maximum working pressures (MWPs) of the pipe and its yield and burst pressures. They show that the explosion pressure would exceed the MWP for the C1D but not for the C1E pipe. For the C1D pipe it would just exceed the yield pressure for the thin wall case, but not the burst pressure for either wall thickness. For the C1E pipe neither yield pressure nor burst pressure was exceeded for either wall thickness. The authors concluded that since the combination of less than nominal wall thickness and loss of corrosion allowance was improbable, it was unlikely that even the C1D pipe would yield, let alone burst. Both C1D and C1E rated pipes should have had no difficulty in containing the explosion.

G.31 For the flanges the Occidental specification was again to ASTM-A106, which requires a yield strength of 36000 psi and a minimum tensile strength of 70000 psi. Sec C of Table G.7 gives the MWP for the 2 ratings of flange. It shows that the explosion would overstress the 1500 lb flange only marginally, but the 900 lb flange severely. Further work would be required to determine what effect this would have. For the stud bolts and nuts it was found that the former would fail before the latter and therefore only the former were considered. The Occidental specifications for the bolts were to ASTM A193-B7. The required material has a yield strength of 105000 psi and a minimum tensile strength of 125000 psi. Sec D of Table G.7 gives for the 2 classes of bolt the bolt thicknesses and the tensile load for yielding. It was assumed that the bolts were initially tight. Sec D also shows the total tensile load resulting from the explosion. It can be seen by comparing the last 2 columns of this section that for both classes 2 bolts are sufficient to prevent yielding. Yielding would take place only if undersized bolts had been used. This is illustrated in Sec E of Table G.7, which shows the combinations of bolt size and number for failure just to occur. If the bolts were slack, twice as many bolts or bolts with diameter larger by the square root of two would be needed to prevent yielding. With regard to the meaning of slackness in this context, Dr Richardson said that it did not matter how tight the bolts were, provided there was no movement. For the ring the Occidental specification was for soft iron octagonal rings. The authors took for these a yield strength of 19100 psi and a tensile strength of 42200 psi. For properly matched flanges, the gap between the raised faces would be 4 mm. The authors found it difficult to envisage failure of the ring in this case. They thought a very small gap might perhaps open up but it would close again when the explosion pressure decayed. If mis-matched flanges were used, the ring would be confined at only one end. Calculations based on the simplification of treating the ring as an infinitely long cylinder gave a large deformation at 200 bara and bursting at 300 bara. The authors found it difficult to predict whether failure would occur in this case. The report also considered bending of the blind flange. If only 2 diametrically opposed bolts were used, it would be possible for the flange to bend about the line joining these bolts. It was estimated that for an explosion of 300 bara a gap 0.1 mm might open up, but it was expected that it would close again when the pressure decayed.

G.32 As far as concerns passage of any flame to the outside, this is determined by the gap available. The parameter generally quoted is the maximum safe gap, which is the maximum width of gap through which a flame will not propagate. The report quoted for lower hydrocarbons a maximum safe gap of less than 1 mm. The equivalent orifice diameter consistent with a gap of this size was 15 mm. Dr Richardson pointed out that the value quoted was for a standard apparatus with a gap length of 1 inch and for atmospheric pressure. The length of the gap on the holes envisaged on the flange assembly would be less and therefore the gap width to just prevent passage of flame would be less; he was unsure of the effect of pressure.

Hydrate formation and behaviour

G.33 The possibility of hydrate formation at various parts of the plant was examined by Drs Richardson and Saville and by Dr Johnsen. The latter also described experimental work on hydrates and discussed hydrate behaviour. An account was given in Chapter 6 of this evidence in so far as it bears on the formation of hydrates at the JT valve and their subsequent behaviour. Further details are given here of the evidence on the formation and behaviour of hydrates both at the JT valve and at other points on the plant.

Hydrate formation

G.34 Evidence on the equilibrium conditions for formation of hydrates at certain critical points was given by Drs Richardson and Saville in their first report and was presented by Dr Richardson. For formation of hydrate to occur it is necessary for the process conditions to lie within a certain envelope of pressure and temperature. If within this envelope the system comes to thermodynamic and phase equilibrium, hydrates will form. The approach to equilibrium may, however, be slow. Thus whether hydrates will actually form depends also on the rate at which equilibrium is approached. There are 2 types of method for the prediction of the equilibrium conditions for hydrate formation. The traditional method is based on K-values for the solid-vapour equilibrium. The other method utilises more fundamental thermodynamics. Drs Richardson and Saville made use of the computer program EQUIPHASE based on the latter method. They estimated the errors in this method as approximately 10-15% in the pressure, plus or minus 3°F in the temperature and 3% w/w in the quantity of methanol required in the aqueous liquid phase. Dr Richardson put the overall error in the amount of methanol to be added to inhibit hydrate formation at about 5% w/w. For example, if the proportion of methanol required was calculated at 28% w/w, the amount needed would lie between 23 and 33% w/w.

Hydrate formation on plant

G.35 The methanol injection points on the plant are shown in Fig J.8. Drs Richardson and Saville calculated first the hydrate formation temperatures assuming no methanol injection. They then calculated the flow rates of methanol required (i) to reduce the hydrate formation temperature to that of the stream exactly and (ii) to reduce it to 5°F below the stream temperature, thus giving a safety margin. The results are shown in Table G.8. From these results the authors concluded that if the methanol injection schedule specified was adhered to, the quantities injected were sufficient to give a safety margin of 5°F between the stream temperature and the hydrate formation temperature. In fact the only point where the margin was as low as 5°F was the JT valve. Additional evidence on the equilibrium conditions for hydrate formation was given by Dr Johnsen. Fig G.2 shows his set of hydrate formation curves for conditions representative of Piper.

Hydrate formation at JT valve

G.36 The possibility of formation of hydrate at the JT valve was considered by Drs Richardson and Saville in their first report on hydrates in general and in their eighth on hydrates at this valve. In their first report, presented by Dr Richardson, they calculated that the hydrate formation temperature just downstream of the JT valve was 62.5°F. They took the temperature at the JT valve as 52.5°F. They estimated that the methanol in the aqueous liquid phase just necessary to prevent hydrate formation at this temperature was 12% w/w. For a temperature of 28-30°F the amount of methanol required in the aqueous liquid phase to inhibit hydrate formation was calculated as 35% w/w. They estimated that given the prescribed methanol injection rates the amount of methanol in that phase would have been 25% w/w. The concentration of methanol would not have been sufficient to prevent hydrate formation.

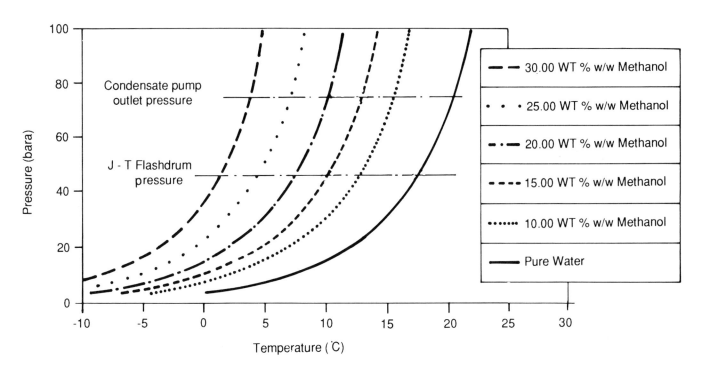

Fig. G.2 Typical hydrate curves for conditions representative of those on Piper given by Dr Johnsen. Hydrates can form if conditions are above and to the left of the relevant curve.

G.37 In their eight report, spoken to by Dr Saville, they presented a graph showing the relation between the hydrate formation temperature and the proportion of methanol in the aqueous liquid layer. At temperatures of 52.5°F and 48.0°F the concentrations of methanol required were 12 and 17% w/w, respectively. At 52.5°F methanol injection rates of 18.0, 19.9 and 37.9 US gal/h gave methanol concentrations in the aqueous liquid phase of 13, 14 and 25% w/w and at 48.0°F methanol concentrations of 14, 16 and 28% w/w. At a temperature of 50.0°F the methanol on the aqueous liquid phase required to prevent hydrate formation was 15% w/w. By interpolation of the above figures, this corresponded to a methanol injection rate at the JT valve of 19.9 US gal/h. This flow corresponds to that to the valve during the interruption of methanol supply.

Hydrate tests

G.38 Dr Johnsen carried out a number of experiments on hydrate formation and behaviour under conditions typical of those on Piper. All the tests were done using a wheel-shaped flow simulator. Condensate was formed in the wheel by admitting a suitable mix of gases and was then brought to equilibrium at the required pressure and temperature by rotating the wheel at 0.5 m/s. Water was then admitted and the behaviour of any hydrates formed was observed. Plate 25(a) shows hydrates formed at the observation window on the wheel. A video of some of the experiments was shown to the Inquiry. One test showed that, for simulated conditions at the JT valve of 639 psia (43.5 bara) pressure, 50.2°F (10.1°C) temperature and 15% w/w methanol in the aqueous phase, hydrates formed rapidly. The conclusions from the tests are described in Chapter 6.

Hydrate behaviour

G.39 Dr Johnsen outlined the conditions most favourable for hydrate formation, given that the system is in the hydrate forming region. If water and gas are simply left to stand, the process of hydrate formation may take weeks. If, however, water is sprayed into gas, near ideal conditions occur for hydrate formation. Hydrates can form long before a stoichiometric equilibrium has formed between the water and gas molecules. In some situations not all the water will be converted to hydrates and water and hydrates will co-exist. In light condensate streams with velocities of less than 1 m/s the water is hardly dispersed at all. It moves along the bottom of a horizontal pipe and gathers in pools in low points in the pipework or ahead of upward pointing bends.

G.40 Hydrates formed in streams with poor pipe-wetting or water-dispersing properties present the most problems, sticking on any mechanical obstructions such as welded joints, pipe seams, tappings, branches, bends and valves or even smooth surfaces. Hydrate plugs form in 2 ways. One mechanism is where sticky hydrates adhere to a wall and provide an anchor for other hydrates as they arrive. The hydrates may then pack and form an ice-like structure. The other is where the hydrates are not sticky but form a soft plug, covering perhaps 10% of the cross-section of the pipe. Upon hitting an obstruction the hydrate "train" will start compacting so that within a few seconds a plug is formed. Hydrates which have adhered to a pipe wall may loosen upon a slight temperature increase and form a soft travelling plug which on hitting an obstruction may compact into a hard plug. Hydrates formed at one point may give rise to problems at another point.

G.41 Once hydrates are formed, their dissolution tends to require quite a lot of heat, and therefore time. The mechanism of heat transfer for dissolution (conductivity) is different from those for formation (convection and turbulence as well as conductivity). The dissolution of hydrates requires considerable amounts of heat. Thus hydrates tend to dissolve only slowly and they may survive for several hours after being transported into parts of the process which have pressures and temperatures outside the hydrate region. Hydrates may be removed by increasing the temperature, adding methanol or decreasing the pressure. If the temperature is raised or methanol is added, this is liable to loosen hydrates from the pipe walls, but not to form water again quickly, so that a hydrate slurry moves down the pipe. This slurry may then compact into a hard plug. If the pressure differential across a hydrate plug is increased and as a result the plug moves, it may travel at the velocity of a rifle bullet. While this is not a likely mode of failure for small pipes, it may be for larger pipes.

Source of ignition

G.42 A brief account was given in Chapter 6 of the review of sources of ignition by Dr J G Marshall. Further details of this review are given here. Ignition of a hydrocarbon vapour cloud may occur either by deposition into the cloud of a small quantity of energy, such as that from a spark, or by heating up the bulk gas until it reaches its autoignition temperature. Dr Marshall emphasised the small amount of energy required to ignite a hydrocarbon vapour cloud. He gave minimum ignition energies of 0.29 mJ for methane and energies in the range 0.24-25 mJ for ethane, propane and butane. The energy needed is therefore very small, equivalent to the energy dissipated by a 5W torch bulb in 50 microseconds. Ignition sources considered by Dr Marshall included electric arcs and sparks, static electricity, flames and hot gases, hot surfaces, hot particles and chemical energy.

Electric arcs and sparks

G.43 Electrical equipment is a potential source of ignition. Equipment for use in hazardous areas is of 2 main types, flameproof and intrinsically safe. The latter is designed to be incapable of giving an incendive spark. Other electrical equipment might in principle give arcs and sparks. It should not do so if correctly selected, installed and maintained, but might if it had suffered damage or deterioration. Mr W N MacLaren, an Electrical Engineer Surveyor with Lloyds Register, described the electrical equipment which would be found in B and C Modules. It included electric motors, instrumentation and controls, junction boxes, cables, light fittings, telephone and public address systems, and trace heating. He had noted in one of his reports a number of bolts missing on an explosion-protected enclosure and other instances of bolts and screws missing, but he considered that the incidence of such missing bolts and screws was low and consistent with that on a well maintained installation. In 1979 a hydrocarbon gas leak occurred in C Module when a valve was inadvertently left open. The source of ignition was identified by the DEn inspector as a junction box.

Electrostatic sparks

G.44 The conditions under which an electrostatic charge builds up and then discharges were reviewed by Dr Marshall. The 2 situations which he highlighted were static discharge from an isolated conductor and from the human body. A jet of vapour containing fine liquid droplets tends to generate an electrostatic charge. A release of liquid condensate from high pressure would give such a jet. If a jet of such vapour impinges on a body which is a conductor but which is insulated from earth, it can cause that body to become charged. If the insulation from earth is sufficiently good, the charge will not dissipate by simply leaking away, but will build up. When a sufficient charge has accumulated, a spark discharge to earth may occur. The spark may have sufficient energy to ignite any flammable gas mixture which is present. Dr Marshall quoted a Shell expert, Mr Strawson, with regard to static discharge from an isolated conductor to the effect that "this mechanism has been responsible for all the explosions known to have resulted from charged liquid sprays". Examples of isolated conductors given by Dr Marshall were the nozzle on the end of a rubber or plastic hose or a spanner not in direct contact with the ground but lying on a piece of rag. The possibility was considered that under certain conditions, the metal scaffolding at the site of PSV 504 might have acted as an isolated conductor; there was no requirement to earth the scaffolding. Dr Marshall thought that the electrical capacity of the scaffolding would probably be too great and the resistance of the path to earth through the wooden planks on which it stood would probably be too low to prevent dissipation of the charge, while if even a high potential did build up the path around the wooden plank to earth would probably be too long to permit incendive discharge.

G.45 Another source of static discharge considered was the human body. Mr Richard had gone into C Module to investigate the first gas alarm and may have entered a flammable cloud. Dr Marshall stated that a prerequisite for the build-up of a significant static charge on a human being would be that he is wearing rubber or other insulating boots; with leather or antistatic footwear any charge would leak rapidly away to the metal deck. According to Mr Richards, the OIM, all personnel, Occidental and contractors, wore leather toe 'tector boots.

Flames and hot gases

G.46 The flare was a large open flame, and was thus the strongest and most obvious source of ignition on the platform, but ignition by this source requires that the flammable gas cloud actually reaches the flare. Hot work activities such as cutting and welding that night were described in Chapter 3. Mr Bollands had no recollection of any hot work permits out that evening for A, B or C Modules and the deluge systems were on automatic, but there was one out for the 68 ft level. There were hot work permits out for the pump room area and an area described as the "habitat" at the east end of D Module. However, construction work would normally finish by 21.00 hours.

G.47 The painters' air compressor, a diesel-driven machine, was situated at the north-west corner of the 68 ft level just to the west of the paint store and, evidently, just outside the boundary of the Zone 2 area of that level. The compressor was used to provide compressed air for the painters, who also did needle-gunning work. In addition, it provided compressed air for the divers' air winches, including that for the divers' bell. According to Mr J P Dixon, a painter, it was the practice for a colleague, Mr S Glendinning, to stay on in the evening to fill up the compressor with diesel, an operation which usually took him from about 18.00 to 21.00 hours. He was fairly sure that Mr Glendinning would not be working as late as 22.00 hours; he was unsure whether it was Mr Glendinning's practice to run the compressor to check it before leaving the job. Mr Punchard, the diving inspection controller, stated that some weeks before he had noticed sparks being emitted from this compressor. He had reported it and believed remedial action had been taken. On the night as he passed by the compressor making his return to the north-west corner it occurred to him that the compressor might act as a source of ignition for any escaping gas and he decided to shut it down. He could not find the shutdown button and asked Mr Elliott, who was passing, to assist him. Mr Elliott stated that it was he who shut the machine down; he had no idea who had asked him to do it. If the painters' air compressor had been running and had been sparking, it could have acted as a source of ignition for any gas cloud at the north-west corner of the 68 ft level or perhaps one falling down from the 84 ft level and being blown north, but there was no evidence that it was sparking on the night of 6 July.

G.48 Both a cigarette and the materials used to light up, such as matches or lighters, are in principle potential sources of ignition. Dr Marshall stated that it is quite difficult to ignite a hydrocarbon cloud with a lit cigarette and that in experimental trials it is usually necessary to do quite a large number of tests before ignition is achieved. Matches and lighters, however, give ignition much more easily. In general, there is a no smoking rule in hazardous areas and the rule is strictly enforced. The rule applied on Piper.

Hot surfaces

G.49 If a flammable mixture is heated up, it eventually ignites spontaneously. The minimum temperature at which this occurs is the autoignition temperature (AIT). Whilst it is convenient for the purposes of science and engineering to utilise the concept of an AIT, this is no more than a way of getting a handle on what are in reality quite complex combustion phenomena. Further, there is no established method of determining the AIT of a mixture of fuel gases. Dr Marshall gave a table of AITs, ranging from 595°C for methane to 285°C for pentanes. A flammable gas mixture can be ignited by a hot surface. In engineering design the temperature at which this occurs is frequently, but conservatively, taken as the AIT. Thus BS 5345 sets the AIT as the maximum for the temperature of a hot surface. In practice, there is a considerable but variable excess of the hot surface ignition temperature over the autoignition temperature. Dr Marshall quoted the statement of Mr Powell that "the case of ignition of gases and vapours bears little or no relation to their ignition temperatures". The lowest excesses shown in a tabulation given by Dr Marshall for small surface areas (not exceeding 108 mm^2, were 302°C for methane, 470°C for propane and 720°C for pentane. The excess also varied with the type of metal constituting the hot surface. Dr Marshall also quoted experiments showing various degrees of temperature excess.

G.50 The casings of the gas turbines on the centrifugal compressors in C Module ran hot. The possibility was explored that such a casing might have been the ignition source. For hot surface ignition to occur due to this source, it would be necessary for a flammable mixture to have access to the surface and for the surface to be sufficiently hot to cause ignition. Mr J Murray, an operator, stated that under normal running conditions the engine casing glowed red, a dull red. One of the checks carried out by the operator was for uneven glow on the engine. Asked to interpret "dull red" in terms of temperature, Dr Marshall referred to a table given by Dr Drysdale, according to which the first visible red glow would be at about 550°C and a dull red at about 700°C. Another operator, Mr J E Cotter, related an incident some 6 or 7 years before in which in the course of his inspection of a turbine he observed a dull red glow from the casing of the machine. He informed Mr Smith, the maintenance lead hand, but was told there was no cause for concern. This evidence suggests it would need experimental work to determine whether in normal operation the temperature of the engine casing might be high enough to cause hot surface ignition of a flammable gas mixture.

G.51 For this to occur, however, it would be necessary for the gas to gain access to the turbine. There are in principle 3 ways in which this might occur. One is for it to enter through an open door, another for it to be drawn in through the ventilation intake and the third for it to be ingested into the turbine air intakes. These 2 intakes were in areas classified as safe, but this does not rule out the presence of gas under more extreme accident conditions. The turbine compartments were maintained under positive pressure by the ventilation fan. There were doors on the compartments which should normally have been closed. The possibility was explored that they might have been open. Several witnesses gave evidence that the doors of both compartments were always kept closed; this was necessary to maintain a proper circulation for cooling. In any case, according to Mr Murray, there was a good air flow, normally sufficient to maintain positive pressure even with the door open.

G.52 Entry of gas through the ventilation system would have required that there was a flammable concentration at the ventilation air intake and that the ventilation system kept running. The ventilation intake was some 11 ft above the deck grating, some 2-3 ft east of the east crane pedestal and therefore projecting outside the module. Dr Davies stated that in his wind tunnel tests he had sensors near the ventilation intake and that though some flammable concentrations might just get in, he thought it unlikely that significant flammable masses would have been ingested. The ventilation system was designed to continue running for 2 hours after compressor shutdown, though if a high alarm were registered on the gas detector at the ventilation system inlet, the system would shut down immediately. The third compressor tripped before the final set of gas alarms. It is likely that the high gas alarm level was not reached in the ventilation intake on any compressor but if it was, the ventilation system should have shut down.

G.53 The combustion air intakes were some 6 or 7 ft up. Again Dr Davies stated that the wind tunnel tests showed that some flammable concentration might just enter. However, there was information from Occidental quoted by Dr Davies to the effect that no more than 25% of the air drawn into these intakes was from the module, the rest being from outboard of it. From Table G.3 the maximum steady-state concentration of gas in the air leaving the module for a leak of 100 kg/min (series 42) is 3.7%. A stream diluted by a factor of 4 would have a concentration of 0.9%, which is below the LEL of 2.1%. Flashback out through the turbine air intakes was not explored nor was entry of gas into these intakes after compressor shutdown.

G.54 If an explosion had occurred within the turbine compartment, the latter would have sustained massive damage. Such damage would probably have been obvious to Captain Morton on the *Maersk Cutter*. Further, this scenario requires that there was then a second explosion in the module itself. The witness evidence is against this.

Hot particles

G.55 Hot particles mentioned by Dr Marshall included hot soot and molten metal from welding and cutting. Sources of hot soot were heaters and engines and the flare. The possibility was raised that closing of a door after inspection of the centrifugal compressor or turbine compartment, such as Mr Richard may have made in seeking to detect the gas leak, might give rise to a mechanical spark of sufficient strength to ignite a flammable gas cloud but insufficiently conspicuous to have been noticed on other occasions. Dr Marshall agreed that with a heavy metal door, shut vigorously and making a glancing contact with the door jamb, this was possible.

Chemical energy

G.56 There are various ways in which energy release by chemical reaction can act as a source of ignition. Dr Marshall referred to 2 in particular, catalytic reactions on gas detector heads and self-heating of oil-soaked lagging or rags. The principle of operation of a gas detector is that the flammable gas is made to react by the catalytic element in the detector and the resultant temperature rise is then measured. In order to guard against ignition from this source, the detector head is equipped with a flame arrester. If the detector head were damaged, protection would be lost.

Probability of ignition and of identifying its source

G.57 Dr Marshall was asked his view of the probability that a gas leak in a module would be ignited. He regarded ignition of a violent release as probable but ignition of a release which was not violent as possible rather than probable.

G.58 At the end of his evidence Dr Marshall was asked how likely he thought it was that it would be possible to identify with reasonable certainty a source of ignition for the leak on Piper. He stated that in his experience of some 10 to 20 explosion investigations the principal guidance is from information available prior to the incident, debris and eye-witness observations, and that only in some one third of cases had he been satisfied that the source of ignition had been found. He agreed that in general in the absence of the debris evidence the chances of identifying the source of ignition were low, that this was so in the case of Piper and that probably the most which could be expected was to narrow the number of possibilities.

Table G.1 - Ventilation rates obtained for C Module on 6 July 1988 in wind tunnel tests on 1: 100 scale model

Wind direction relative to platform north(°)	Ventilation rate (m³/s)		
	Wind speed (m/s)		
	7.2	8.2	9.2
200	31	35	39
207	40	46	52
218	51	58	65
228	57	65	73

Table G.2 - Complete list of wind tunnel tests conducted

Series	Location	Leak Type	Angle (°)	Rate (kg/min)	Ingestion & exhaust In (cfm)	Location	Out (cfm)
First set							
10	1	J	150	50	—	—	—
11	1	J	120	50	—	—	—
12	1	C	—	50	—	—	—
13	1	C	—	50	—	—	6000
14	1	C	—	50	19000	C	—
15	1	C	—	50	19000	C	6000
16	1	J	150	50	19000	C	6000
17	1	J	150	50	6800	C	6000
18	1	J	120	50	6800	C	6000
19	1	J	90	50	6800	C	6000
20	1	J	150	50	6800	A	6000
21	1	J	150	25	6800	C	6000
22	1	J	90	25	6800	C	6000
23	1	J	60	25	6800	C	6000
24	1	J	90	37	6800	C	6000
25	1	J	150	37	6800	C	6000
26	1	J	180	37	6800	C	6000
27	1	J*	180	37	6800	C	6000
28	1	C,2	180	37	6800	C	6000
29	1	C,1	120	37	6800	C	6000
30	1	C,1	120	37	—	—	6000
31	1	C,1	120	37	6800	A	6000
32	1	C,1	120	37	18200	C	6000
33	1	C,1	120	37	7000	C	6000
34	1	C	—	12	7000	C	6000
35	1	C,1	120	10	7000	C	6000
36	1	C,1	120	4	7000	C	6000
37 Aborted							
38	1	C,1	120	1	7000	C	6000
39	1	C,1	90	10	7000	C	6000
40	1	C,1	150	10	7000	C	6000
41	1	J	150	10	7000	C	6000
42	1	C,1	120	100	7000	C	6000
43	2	C,1	120	37	7000	C	6000
44	3	C	—	50	7000	C	6000
Second set							
45	1	J	150	100	7000	C	6000
46	2	J	150	100	7000	C	6000
47	3	J	150	100	7000	C	6000
48	4	J	150	100	7000	C	6000
49	1	J	150	1	7000	C	6000
50	2	J	150	1	7000	C	6000
51	3	J	150	1	7000	C	6000
52	1	P	—	100	7000	C	6000

Notes:

Series: Group of runs at the same conditions.

The following applies to the first set of tests (series 10-44):

Type: J   — jet (small hole in horizontal pipe)

       J*   — jet impinging on nearby plate (separation 1m)

       C   — circumferential leak

       C,n — circumferential leak with n x 120° sectors open

Angle: Orientation of jet

       0° vertically upward

       90° horizontal to east

       180° vertically downward, etc.

Gas: All series used propane at -42°C except series 44 which used a cold natural gas, ethane and propane mixture

Ventilation: Wind induced ventilation was 46 m³/s in all series

Ingestion: Volume of air drawn from within C Module

Location: intake of A or C centrifugal compressor

Exhaust: Total exhaust volume of vent air from all 3 centrifugal compressors

The following applies to the second set of tests (series 45-52):

Type: P — Release from an open horizontal pipe 3½ inch in diameter directed towards south wall

Gas: All series a cold natural gas, ethane and propane mixture

Otherwise as for first set tests.

Table G.3 - Concentrations and times to alarm obtained for selected wind tunnel tests: first set

		Series								
		15	16	19	26	27	32	35	36	42
Leak rate (kg/min)		50	50	50	37	37	37	10	4	100
Leak type		C	J	J	J	J$\star$	C,1	C,1	C,1	C,1
Leak angle		—	150	90	180	180	120	120	120	120

A - Steady-state concentrations (% v/v)

Sensor	Area	15	16	19	26	27	32	35	36	42
G101/1	C2	1.7	0	0	0	0.6	0.2	0	0	0
G101/2	C2	1.6	2.3	2.5	2.2	2.5	2.0	0.9	0.8	3.1
G101/3	C2	1.2	1.1	0	0	1.1	0	0.2	0	0.3
G103/1	C3	2.7	3.4	2.6	2.7	1.5	2.4	1.2	1.2	3.7
G103/2	C4	1.6	2.5	x[a]	2.4	x	1.7	0.9	0.8	1.9
G103/3	C5	1.0	0.7	x	1.5	x	0.9	0.3	0.3	2.2
G27	C3	2.2	1.8	2.9	1.1	1.6	2.8	0.5	0.5	3.3
G30	C4	1.5	2.8	x	1.4	x	1.9	0.6	0.7	1.8
G33	C5	1.0	0.8	x	0.6	x	0.7	0.2	0	2.0

B - Time to low level alarm(s)

Sensor	Area	15	16	19	26	27	32	35	36	42
G101/1	C2	5	—	x	—	15	—	—	—	—
G101/2	C2	40	40	25	45	30	40	60	65	15
G101/3	C2	45	25	x	—	30	—	—	—	—
G103/1	C3	15	15	10	30	55	15	30	30	5
G103/2	C4	60	20	x	30	x	30	70	85	20
G103/3	C5	105	120	x	55	x	55	—	—	30
G27	C3	15	15	15	40	45	15	85	80	10
G30	C4	70	15	x	60	x	25	85	80	25
G33	C5	110	100	x	75	x	120	—	—	25

C - Time to high level alarm(s)

Sensor	Area	15	16	19	26	27	32	35	36	42
G101/1	C2	—	—	x	—	—	—	—	—	—
G101/2	C2	—	—	140	85	140	—	—	—	40
G101/3	C2	—	—	x	—	—	—	—	—	—
G103/1	C3	125	35	25	75	—	—	—	—	10
G103/2	C4	—	180	x	155	x	—	—	—	—
G103/3	C5	—	—	x	—	x	—	—	—	—
G27	C3	80	—	40	—	—	30	—	—	20
G30	C4	—	55	x	—	x	—	—	—	—
G33	C5	—	—	x	—	x	—	—	—	—

D - Area showing first alarm

		15	16	19	26	27	32	35	36	42
		C2	C3/C4	C3?	C3/C4	C2?	C3	C3	C3	C3

E - Time from first to second area alarm(s)

		15	16	19	26	27	32	35	36	42
		10	0	?	0	15?	10	30	35	10

Notes:

(a) x indicates that there was no sensor at this point in this series.

(b) For key to leak type and leak angle see notes to Table G.2.

Table G.4 - Concentrations and times to alarm obtained for selected wind tunnel tests: second set

	Series						
	43[a]	44	45	46	47	48	52
Gas	P	NG	NG	NG	NG	NG	NG
Position	2	3	1	2	3	4	1
Leak rate (kg/min)	37	50	100	100	100	100	100
Leak type	C,1	C	J	J	J	J	P
Leak angle	120	—	150	150	150	150	—

A - Steady-state concentrations (% v/v)

Sensor	Area							
G101/1	C2	0	0.6	0.5	0	0	5.1	5.5
G101/2	C2	3.0	2.5	2.6	4.2	3.1	1.6	2.6
G101/3	C2	0	1.9	3.7	4.7	4.5	3.1	2.8
G103/1	C3	2.0	2.5	3.7	3.1	4.8	4.2	4.0
G103/2	C4	2.0	2.5	3.4	3.0	3.1	2.9	2.9
G103/3	C5	1.9	1.0	2.0	3.5	1.3	1.4	1.1
G27	C3	1.4	1.8	3.5	3.3	5.1	4.3	3.2
G30	C4	1.8	3.5	2.8	2.4	3.4	3.2	2.9
G33	C5	1.4	1.7	2.4	4.0	1.4	1.4	1.4

B - Time to low level alarm(s)

Sensor	Area							
G101/1	C2	—	115	—	—	—	31	7
G101/2	C2	10	30	35	2	19	84	61
G101/3	C2	—	50	22	5	12	61	25
G103/1	C3	30	40	15	22	35	65	38
G103/2	C4	25	35	14	24	24	76	58
G103/3	C5	25	60	36	16	32	79	112
G27	C3	30	45	22	22	21	73	34
G30	C4	25	30	35	38	31	32	67
G33	C5	25	30	19	11	31	76	77

C - Time to high level alarm(s)

Sensor	Area							
G101/1	C2	—	—	—	—	—	59	18
G101/2	C2	35	120	—	25	—	—	—
G101/3	C2	—	—	—	11	36	—	—
G103/1	C3	—	180	—	—	46	79	118
G103/2	C4	—	120	—	—	—	—	—
G103/3	C5	—	—	—	—	—	—	—
G27	C3	—	—	—	—	33	121	—
G30	C4	—	180	—	—	—	—	—
G33	C5	—	—	—	35	—	—	—

D - Area showing first alarm

C2	C2,C4,C5	C4	C2	C2	C2	C2

E - Time from first to second area alarm(s)

15	0	1	9	9	1	27

Notes:

(a) For convenience series 43 and 44, which are for locations other than position 1, are included in this table rather than in Table G.3.

(b) Gas: P = propane; NG = natural gas

(c) For key to position, leak type and leak angle see notes to Table G.2.

Table G.5 - Compositions of process streams used in the FLACS code explosion simulations

Component	Composition (% v/v)	
	Natural gas	Condensate
A		
Methane (C1)	67.75	19.86
Ethane	15.76	18.98
Propane (C3)	9.34	31.06
Butanes	1.97	17.16
Pentanes	0.45	9.94
Carbon dioxide	0.45	0.27
Nitrogen	4.23	0.46
B		
Equivalent C1	87.66	40.04
Equivalent C3	12.34	59.96

Table G.6 - Work on PSVs 504 and 505 in the period 1980-1988 spoken to by witnesses

Date	PSV	Test certificate	Flange rating (lb)	Signature for valve overhaul
		No		
9.4.1980	505	-	None	Whalley
30.4.1980	504	-	None	Black
17.9.1981	505	-	None	Smith
7.6.1984	505	3046	None	Cowie
8.6.1984	504	3045	None	Cowie
22.6.1984	504	3049	900	Reid
15.9.1985	504	0018	None	Thom
19.9.1985	505	0033	None	Ritchie
29.10.1985[a]	504	3825	1500	Pirie
29.10.1985[a]	505	3826	1500	Pirie
16.9.1986	504	0101	1500	Barclay
17.9.1986	505	0104	1500	McDonald
24.3.1988	505	2607	1500	Bruce

Note:

(a) This work was resetting of the cold test pressure, which was done with the PSV on the scaffolding, and did not involve fitting a blind flange.

Table G.7 - Conditions for failure of condensate injection pump A pipework, blind flange and flange bolts under explosion pressure

A - Pipework: maximum working pressures

Rating	Pipe schedule	MWP (bara)
C1D	120	180
C1E	XXS	320

B - Pipework: yield and burst pressures

Rating	Yield pressure (bara)		Burst pressure (bara)	
	Nominal	Minimum wall	Nominal	Minimum wall
C1D	420	280	890	540

C - Flange

Rating	Class	MWP (bara)
C1D	900 lb	153
C1E	1500 lb	255

D - Bolts: tensile loads on bolts

Rating	Bolt diameter (in)	Tensile load for yielding per bolt (lbf)	Total tensile load for explosion pressure of 300 bara (lbf)
900 lb	1.125	83000	116000
1500 lb	1.250	105000	139000

E - Bolts: bolt diameters and numbers for failure just to occur

Rating	Bolt diameter (in)		
	2 bolts	4 bolts	8 bolts
900 lb	0.875	0.625	0.437
1500 lb	1.000	0.625	0.500

Table G.8 - Hydrate formation temperatures and methanol concentrations to prevent hydrate formation for selected streams in phase 1 operation

Stream	200	210	211	330	340	350
Pressure (psia)	635	635	635	635	670	1100
Temperature (°F)	52.5	55.6	55.6	55.6	55.9	60.2
Hydrate formation temperature (HFT) (°F)	62.5	62.4	62.0	>55.6	>55.9	>60.2
Methanol concentration [a] (% w/w)						
(a) to reduce HFT to stream temperature	12	8	8	5	6	1
(b) to reduce HFT to 5°F below stream temperature	17	14	14	11	11	7
Methanol concentration [a,b] from specified injection rates (% w/w)	17	28	28	28	28	36

Notes:

(a)   In aqueous liquid phase.

(b)   Calculated from the methanol injection rates specified by Mrs Paterson.

Appendix H

Schedule of Information Relating to the Deceased

Deceased	Occupation	Employer	Recovery/LKW	Principal Cause of Death
1. ADAMS, ROBERT McINTOSH (39) 70 Rowan Road Aberdeen	Rigger	John Wood Group plc John Wood House Greenwell Road East Tullos Aberdeen	Recovered 26.10.88 in D Deck of ERQ at Flotta	Inhalation of smoke and gas
2. ANDERSON, GEORGE ALEXANDER J (45) 12 Dundonnie Street Boddam Peterhead	Baker	Kelvin Catering Camps Ltd 5 Queen's Terrace Aberdeen	Recovered 26.10.88 in D Deck of ERQ at Flotta	Inhalation of smoke and gas
3. ANDERSON, IAN GEDDES (29) 39 Dunnydeer Park Insch	Dual Service Operator	Halliburton Manufacturing & Services Ltd Howe Moss Crescent Dyce Aberdeen	Recovered 26.10.88 in D Deck of ERQ at Flotta	Inhalation of smoke and gas
4. ANDERSON, JOHN (33) 4 Dunure Place Kirkcaldy Fife	Catering Manager	Kelvin Catering Camps Ltd 5 Queen's Terrace Aberdeen	Recovered 7.7.88 on surface of sea near Platform	Inhalation of smoke and gas
5. ASHTON, MARK DAVID (19) 23 Townhead Drive Inverurie	Trainee Technician/Cleaner	Macnamee Services Ltd Burnside Works Kennerty Mills Road Peterculter Aberdeen	Recovered 24.10.88 in D Deck of ERQ at Flotta	Inhalation of smoke and gas
6. BAIN, WILSON CRAWFORD A (34) 8 Pineview Fraserburgh	Valve Technician*	John Wood Group plc John Wood House Greenwell Road East Tullos Aberdeen	Missing: LKW – in Accom. between 22.00 and 22.20	N/A
7. BARBER, BARRY CHARLES (46) Annanbank Lewis Street Stranraer	Diving Consultant*#	Aberdeen Offshore Services Ltd Union Buildings 15 Union Street Aberdeen	Recovered 10.11.88 Track 8 by *Heather Sprig*: LKW – in water between B3 and B4 Legs shortly after 22.20	Drowning

	Deceased	Occupation	Employer	Recovery/LKW	Principal Cause of Death
8.	BARCLAY, CRAIG ALEXANDER (24) 11 Park Avenue Dundee	Welder	John Wood Group plc John Wood House Greenwell Road East Tullos Aberdeen	Recovered 26.10.88 in D Deck of ERQ at Flotta	Inhalation of smoke and gas
9.	BARR, ALAN (37) 20 Carronvale Road Larbert	Electrical Technician	Occidental Petroleum (Caledonia) Ltd 1 Claymore Drive Bridge of Don Aberdeen	Missing	N/A
10.	BATCHELOR, BRIAN PHILIP (44) 2 Allen Close Fleetwood	Seaman	Haven Shipping Co. Queens Road Great Yarmouth	Recovered 7.7.88 on surface of sea near Platform: LKW – in FRC of *Sandhaven* on W side of Platform about 22.50	Drowning
11.	BORG, AMABILE ALEXANDER (51) "Ithaca" The Drive Bassleton Lane Thornaby Cleveland	Non-Destructive Tester	Testwell (Ultrasonics) Ltd Unit 27 Murcar Industrial Est. Bridge of Don Aberdeen	Recovered 24.10.88 in D Deck of ERQ at Flotta	Inhalation of smoke and gas
12.	BRACKENRIDGE, HUGH WALLACE (47) 34 Kirk Street Leith Edinburgh	Roustabout	Bawden International Ltd Wellheads Road Dyce Aberdeen	Missing	N/A
13.	BREMNER, ALEXANDER ROSS COLVIN (38) "San Mar" 2 Edward Avenue Banff	Production Operator★	Occidental Petroleum (Caledonia) Ltd 1 Claymore Drive Bridge of Don Aberdeen	Missing: LKW – on NW Navigation Aid Platform shortly before 22.20	N/A
14.	BRIANCHON, ERIC ROLAND PAUL (32) Rue de L'Oiseau Bleu Saint Wandrille Rancon France	Technician	Coflexip SA Rue Jean Hure Le Trait France 76580	Died in hospital on 19.7.88	Extensive Burns

	Deceased	Occupation	Employer	Recovery/LKW	Principal Cause of Death
15.	BRISTON, HUGH (40) 7 Curran Avenue Whinney Banks Middlesbrough	Scaffolder	Aberdeen Scaffolding Company Ltd Harness Road Aberdeen	Recovered 6.9.88 in Galley of ERQ	Inhalation of smoke and gas
16.	BROWN, HENRY (39) 11 Derwent Drive Coatbridge	Welder	John Wood Group plc John Wood House Greenwell Road East Tullos Aberdeen	Recovered 25.10.88 in D Deck of ERQ at Flotta	Inhalation of smoke and gas
17.	BROWN, STEPHEN (27) 73 Dame Dorothy Crescent Roker Sunderland	Assistant Chef/Baker	Kelvin Catering Camps Ltd 5 Queen's Terrace Aberdeen	Recovered 26.10.88 in D Deck of ERQ at Flotta	Inhalation of smoke and gas
18.	BRUCE, GORDON CRAIB (51) 55 Great Northern Road Aberdeen	Helicopter Landing Officer	John Wood Group plc John Wood House Greenwell Road East Tullos Aberdeen	Recovered 24.10.88 in D Deck of ERQ at Flotta	Inhalation of smoke and gas
19.	BRUCE, JAMES (42) 12 Beech Road Stockethill Aberdeen	Logger	John Wood Group plc John Wood House Greenwell Road East Tullos Aberdeen	Recovered 11.7.88 50m, 245 degrees from Leg A5: LKW – Climbing up ladder to Drilling Tea Hut about 22.15	Inhalation of smoke and gas
20.	BUSSE, CARL WILLIAM (31) 408 Johnson Navasota Texas 77868 USA	Directional Drilling Supervisor	Eastman Christensen Ltd Murcar Commercial Pk Denmore Road Bridge of Don Aberdeen	Recovered 24.10.88 in D Deck of ERQ at Flotta	Inhalation of smoke and gas
21.	CAMPBELL, DAVID (23) 1 Hill Street Lossiemouth Morayshire	Cleaner	Macnamee Services Ltd Burnside Works Kennerty Mills Road Peterculter Aberdeen	Recovered 23.10.88 in D Deck of ERQ at Flotta	Inhalation of smoke and gas
22.	CAMPBELL, DAVID ALLEN (29) 2 Ash Grove The Hillock Portlethen Aberdeen	Scaffolder	Aberdeen Scaffolding Company Ltd Hareness Road Aberdeen	Recovered 7.7.88 on surface of sea near Platform: LKW – jumping off Helideck about 22.50	Inhalation of smoke and gas

Deceased	Occupation	Employer	Recovery/LKW	Principal Cause of Death
23. CARGILL, ALEXANDER WATT (39) 17 Howard Street Arbroath	Electrician	John Wood Group plc John Wood House Greenwell Road East Tullos Aberdeen	Recovered 24.10.88 in D Deck of ERQ at Flotta	Inhalation of smoke and gas
24. CARROLL, ROBERT (34) 40 Coronation Road Peterculter Aberdeen	Safety Operator*	London Bridge Eng Ltd Aberdeen Business Centre Willowbank House Willowbank Road Aberdeen	Missing: LKW – on NW Navigation Aid Platform shortly before 22.20	N/A
25. CARTER, ALAN (43) 30 Dixons Bank Marton Middlesbrough	Lead Production Operator	Occidental Petroleum (Caledonia) Ltd 1 Claymore Drive Bridge of Don Aberdeen	Recovered 24.10.88 in D Deck of ERQ at Flotta	Inhalation of smoke and gas
26. CLELAND, ROBERT (33) 89 Cairngorm Gardens Eastfield Cumbernauld	Derrickman	Bawden International Ltd Wellheads Road Dyce Aberdeen	Recovered 10.7.88 GMR 340062E 6484135N	Inhalation of smoke and gas
27. COLE, STEPHEN COLIN (40) 10 Milson Road Keelby Lincolnshire	Radio Officer	Inspectorate UK plc Unit 3 Wellheads Way Dyce Aberdeen	Recovered 7.7.88 on surface of sea near Platform: LKW – possibly at entrance to Offices from pipe deck after 22.20	Inhalation of smoke and gas
28. CONNOR, HUGH (35) 6 Corse Street West Kilbride	Instrument Technician/Lecturer	John Wood Group plc John Wood House Greenwell Road East Tullos Aberdeen	Recovered 7.7.88 on surface of sea near Platform: LKW – leaving Instrument Workshop at about 22.00	Drowning
29. COOKE, JOHN EDWARD SHERRY (59) 18 Castle Road Greenock	Plater	John Wood Group plc John Wood House Greenwell Road East Tullos Aberdeen	Missing	N/A
30. COOPER, JOHN THOMAS (37) 6 Haswell Avenue Hartlepool	Instrument Technician	John Wood Group plc John Wood House Greenwell Road East Tullos Aberdeen	Missing: LKW – leaving Instrument Workshop at about 22.00	N/A

	Deceased	Occupation	Employer	Recovery/LKW	Principal Cause of Death
31.	COUTTS, WILLIAM NUNN (37) 2 Morangie Road Tain	Chef	Kelvin Catering Camps Ltd 5 Queen's Terrace Aberdeen	Recovered 26.10.88 in D Deck of ERQ at Flotta	Inhalation of smoke and gas
32.	COWIE, WILLIAM JOHN (32) "Trevear" South Pringle Street Buckie	Steward	Kelvin Catering Camps Ltd 5 Queen's Terrace Aberdeen	Recovered 25.10.88 in D Deck of ERQ at Flotta	Inhalation of smoke and gas
33.	COX, MICHAEL JOHN (26) 43 Pentland Crescent Aberdeen	Scaffolder	Aberdeen Scaffolding Company Ltd Hareness Road Aberdeen	Recovered 17.7.88 130m, 030 degrees from Leg A.5: LKW – on top of Radio Room at about 22.35	Drowning
34.	CRADDOCK, ALAN IRVIN (31) 10 Devanha Terrace Aberdeen	Drilling Supervisor#	Occidental Petroleum (Caledonia) Ltd 1 Claymore Drive Bridge of Don Aberdeen	Recovered 24.10.88 in D Deck of ERQ at Flotta	Inhalation of smoke and gas
35.	CROWDEN, EDWARD JOHN (47) 100 Thistle Drive Portlethen Aberdeen	Electrical Technician★	Occidental Petroleum (Caledonia) Ltd 1 Claymore Drive Bridge of Don Aberdeen	Recovered 24.10.88 in D Deck of ERQ at Flotta	Inhalation of smoke and gas
36.	CURTIS, BERNARD (45) 21 Tayside Street Carnoustie	Deputy Production Supt#	Occidental Petroleum (Caledonia) Ltd 1 Claymore Drive Bridge of Don Aberdeen	Recovered 26.10.88 in D Deck of ERQ at Flotta	Inhalation of smoke and gas
37.	DA SILVA, JOSE HIPOLITO (26) 40 Invercauld Place Aberdeen	Steward	Kelvin Catering Camps Ltd 5 Queen's Terrace Aberdeen	Recovered 24.10.88 in D Deck of ERQ at Flotta	Inhalation of smoke and gas
38.	DAWSON, JOHN STEPHEN (38) 7 Bruce Walk Nigg Aberdeen	Telecom Engineer	Inspectorate UK plc Unit 3 Wellheads Way Dyce Aberdeen	Recovered 22.11.88 Track 19 by *Janeen*: LKW – at N doors of Galley after 22.00	Not Ascertained

467

	Deceased	Occupation	Employer	Recovery/LKW	Principal Cause of Death
39.	DEVERELL, ERIC (51) "Walescot" Bruntland Road Portlethen Aberdeen	Production Clerk	Occidental Petroleum (Caledonia) Ltd 1 Claymore Drive Bridge of Don Aberdeen	Recovered 15.8.88 GMR 340023E, 6483933N	Death in fire
40.	DUNCAN, ALEXANDER (51) 21 Dalmahoy Drive Dundee	Steward	Kelvin Catering Camps Ltd 5 Queen's Terrace Aberdeen	Recovered 7.7.88 on surface of sea near Platform: LKW – outside Galley near ladder to Helideck about 22.40	Chest and abdominal injuries
41.	DUNCAN, CHARLES EDWARD (29) 31 Cairnwell Drive Aberdeen	Floorman*	Bawden International Ltd Wellheads Road Dyce Aberdeen	Recovered 31.10.88 in A Deck of ERQ at Flotta	Inhalation of smoke and gas
42.	DUNCAN, ERIC (49) 10 Dales Road Peterhead	Drilling Materials Man*	Occidental Petroleum (Caledonia) Ltd 1 Claymore Drive Bridge of Don Aberdeen	Recovered 7.11.88 GMR 340011E, 6484145N: LKW – trapped in Materials Room after 22.00	Inhalation of smoke and gas
43.	DUNCAN, JOHN (33) (FORMERLY WALKER) 26 Balgownie Drive Bridge of Don Aberdeen	Engineer	Northern Industrial & Marine Services Company Ltd Unit 15 Woodlands Road Dyce Aberdeen	Recovered 18.7.88 144m, 011 degrees from Leg A5	Drowning
44.	DUNCAN, THOMAS IRVINE (39) 40 Langdykes Drive Cove Bay Aberdeen	Roustabout*	Bawden International Ltd Wellheads Road Dyce Aberdeen	Recovered 14.7.88 84m, 260 degrees from Leg B5: LKW – possibly in Mechanical Workshop after 22.00	Inhalation of smoke and gas
45.	DUNCAN, WILLIAM DAVID (38) 106 Hawick Drive Dundee	Crane Operator	Bawden International Ltd Wellheads Road Dyce Aberdeen	Recovered 20.7.88 205m, 054 degrees from Leg A5: LKW – in corridor of-C Deck after 22.00	Abdominal injury
46.	ELLIS, DAVID ALAN (28) 14 Carrhouse Lane Moreton Wirral Merseyside	Steward	Kelvin Catering Camps Ltd 5 Queen's Terrace Aberdeen	Recovered 25.10.88 in D Deck of ERQ at Flotta	Inhalation of smoke and gas

	Deceased	Occupation	Employer	Recovery/LKW	Principal Cause of Death
47.	FINDLAY, DOUGLAS NEWLANDS (38) Errol Cottage Blackhills Elgin	Supervisor Mechanic	Bawden International Ltd Wellheads Road Dyce Aberdeen	Recovered 26.10.88 in D Deck of ERQ at Flotta	Inhalation of smoke and gas
48.	FLOOK, HAROLD EDWARD GEORGE (51) 2 Fotheringham Drive Monifieth Dundee	Production Operator	Occidental Petroleum (Caledonia) Ltd 1 Claymore Drive Bridge of Don Aberdeen	Recovered 24.10.88 in D Deck of ERQ at Flotta	Inhalation of smoke and gas
49.	FOWLER, GEORGE (40) 14 Kilnwick Close Owington Farm Billingham Cleveland	Electrical Technician	Occidental Petroleum (Caledonia) Ltd 1 Claymore Drive Bridge of Don Aberdeen	Recovered 7.7.88 on surface of sea near Platform	Inhalation of smoke and gas
50.	FREW, ALEXANDER PARK (41) 1 Linn Road Ardrossan Ayrshire	Plater	John Wood Group plc John Wood House Greenwell Road East Tullos Aberdeen	Recovered 25.10.88 in D Deck of ERQ at Flotta	Inhalation of smoke and gas
51.	GALLACHER, SAMUEL QUEEN (30) 29 Regent Street Greenock	Pipe Fitter	John Wood Group plc John Wood House Greenwell Road East Tullos Aberdeen	Recovered 24.10.88 in D Deck of ERQ at Flotta	Inhalation of smoke and gas
52.	GALVEZ-ESTEVEZ, MIGUEL (36) 7 Braehead Drive Cruden Bay Peterhead	Assistant Chef	Kelvin Catering Camps Ltd 5 Queen's Terrace Aberdeen	Recovered 24.10.88 in D Deck of ERQ at Flotta	Inhalation of smoke and gas
53.	GIBSON, ERNEST (45) 498 Hylton Road Sunderland	Mud Engineer	Magcobar Imco GB Ltd Pocra Quay Aberdeen	Recovered 7.7.88 on surface of sea near Platform: LKW – on S side of pipe deck after 22.20	Drowning
54.	GILL, ALBERT STUART (32) 10 Swannay Road Aberdeen	Roustabout	John Wood Group plc John Wood House Greenwell Road East Tullos Aberdeen	Recovered 6.9.88 in Galley of ERQ	Inhalation of smoke and gas

	Deceased	Occupation	Employer	Recovery/LKW	Principal Cause of Death
55.	GILLANDERS, IAN (50) 21 Lodgehill Park Nairn	Instrument Pipe Fitter	John Wood Group plc John Wood House Greenwell Road East Tullos Aberdeen	Missing: LKW – going towards SW corner of Platform shortly after 22.40	N/A
56.	GILLIGAN, KEVIN BARRY (35) 6 Philip Grove Sutton St Helens Merseyside	Steward	Kelvin Catering Camps Ltd 5 Queen's Terrace Aberdeen	Recovered 12.7.88 GMF 340090E, 6484252N	Head and chest injuries
57.	GLENDINNING, SHAUN (24) 40 Hillview Brechin	Painter	John Wood Group plc John Wood House Greenwell Road East Tullos Aberdeen	Recovered 25.10.88 in D Deck of ERQ at Flotta	Inhalation of smoke and gas
58.	GOLDTHORP, JOHN EDWARD THOMAS (37) 10 Craigmill Gardens Carnoustie	Motorman*	Bawden International Ltd Wellheads Road Dyce Aberdeen	Missing: LKW – in Mechanical Workshop after 22.00	N/A
59.	GOODWIN, STEPHEN ROBERT (22) 112 Fitzwilliam Street Swinton Rotherham South Yorkshire	Geologist	Exploration Logging North Sea Ltd Walton Road Dyce Aberdeen	Recovered 5.11.88 in B Deck of ERQ at Flotta	Inhalation of smoke and gas
60.	GORDON, JAMES EDWARD GRAY (38) 72 Thistle Court Aberdeen	Floorman*	Bawden International Ltd Wellheads Road Dyce Aberdeen	Missing: LKW – on Drill Floor after 22.00	N/A
61.	GORMAN, DAVID LEE (41) 7 Forth Wynd Port Seton East Lothian	Safety Operator	London Bridge Eng Ltd Aberdeen Business Centre Willowbank House Willowbank Road Aberdeen	Missing: LKW – in Reception after 22.00	N/A
62.	GRAHAM, KENNETH (40) 2 Pevensey Close Preston Grange North Shields	Mechanical Technician	Occidental Petroleum (Caledonia) Ltd 1 Claymore Drive Bridge of Don Aberdeen	Missing: LKW – on pipe deck shortly before 22.20	N/A

	Deceased	Occupation	Employer	Recovery/LKW	Principal Cause of Death ·
63.	GRANT, PETER JOHN (31) Naeview 20 Drumsmittal Road North Kessock Inverness	Production Operator	Occidental Petroleum (Caledonia) Ltd 1 Claymore Drive Bridge of Don Aberdeen	Recovered 23.10.88 in D Deck of ERQ at Flotta	Inhalation of smoke and gas
64.	GRAY, CYRIL JAMES (49) 171 Heathryfold Circle Aberdeen	Safety Operator	Occidental Petroleum (Caledonia) Ltd 1 Claymore Drive Bridge of Don Aberdeen	Recovered 7.7.88 on surface of sea near Platform: LKW – on S side of pipe deck after 22.20	Drowning
65.	GREEN, HAROLD EUGENE JOSEPH (44) 2 Newburn Court South Shields	Rigger	John Wood Group plc John Wood House Greenwell Road East Tullos Aberdeen	Missing: LKW – on top of Accom. W shortly before 22.20	N/A
66.	GROVES, MICHAEL JOHN (44) 10 Maidlands Edinburgh Road Linlithgow West Lothian	Production Operator	Occidental Petroleum (Caledonia) Ltd 1 Claymore Drive Bridge of Don Aberdeen	Recovered 18.7.88 126m, 076 degrees from Leg A5	Inhalation of smoke and gas
67.	HACKETT, JOHN (49) 7 Brook Road Lymm Warrington Cheshire	Electrical Technician★	Occidental Petroleum (Caledonia) Ltd 1 Claymore Drive Bridge of Don Aberdeen	Recovered 7.7.88 on surface of sea near Platform: LKW – at Reception about 22.05	Inhalation of smoke and gas
68.	HAY, IAN (31) 25 Boston Close Battlehill Wallsend Tyne and Wear	Steward	Kelvin Catering Camps Ltd 5 Queen's Terrace Aberdeen	Recovered 17.7.88 92m, 03 degrees from Leg A5	Inhalation of smoke and gas
69.	HAYES, THOMAS ALBERT (39) 223 Mosslands Drive Wallasey	Rigging Supervisor	John Wood Group plc John Wood House Greenwell Road East Tullos Aberdeen	Recovered 26.10.88 in D Deck of ERQ at Flotta	Inhalation of smoke and gas

	Deceased	Occupation	Employer	Recovery/LKW	Principal Cause of Death
70.	HEGGIE, JAMES (45) 18 Viewforth Gardens Kirkcaldy Fife	Project Services Supt.#	Occidental Petroleum (Caledonia) Ltd 1 Claymore Drive Bridge of Don Aberdeen	Recovered 24.10.88 in D Deck of ERQ at Flotta	Inhalation of smoke and gas
71.	HENDERSON, DAVID WILLIAM (28) 42 Hazlehead Place Aberdeen	Lead Floorman	Bawden International Ltd Wellheads Road Dyce Aberdeen	Recovered 25.10.88 in D Deck of ERQ at Flotta	Inhalation of smoke and gas
72.	HOUSTON, PHILIP ROBERT (35) 4 Dixon Terrace Pitlochry Perthshire	Geologist	NL Sperry Sun/MWD/Baroid Logging Systems Wellheads Road Dyce Aberdeen	Recovered 25.9.88 GMR 340342E, 6484113N	Inhalation of smoke and gas
73.	JENNINGS, DUNCAN (28) 42 Drewett Close Shinfield Reading Berkshire	Geologist	Exploration Logging North Sea Ltd Walton Road Dyce Aberdeen	Recovered 16.7.88 66m, 211 degrees from Leg A5	Drowning
74.	JONES, JEFFREY GRANT (37) 2 Heol Brynhyfryd Woodlands Park Llantwit Fardre	Assistant Driller*	Bawden International Ltd Wellheads Road Dyce Aberdeen	Missing: LKW – on pipe deck after 22.20	N/A
75.	KAVANAGH, CHRISTOPHER (49) 71 Lomand Road Wemyss Bay	Plater	John Wood Group plc John Wood House Greenwell Road East Tullos Aberdeen	Recovered 29.7.88 116m, 327 degrees from Leg A5	Multiple injuries
76.	KELLY, WILLIAM HOWAT (43) 10 Summerfield Road Condorrat Cumbernauld	Electrical Technician	John Wood Group plc John Wood House Greenwell Road East Tullos Aberdeen	Recovered 25.10.88 in D Deck of ERQ at Flotta	Inhalation of smoke and gas

	Deceased	Occupation	Employer	Recovery/LKW	Principal Cause of Death
77.	KILLINGTON, IAN (33) 5 Fulmar Drive Blyth Northumberland	Steward	Kelvin Catering Camps Ltd 5 Queen's Terrace Aberdeen	Recovered 25.10.88 in D Deck of ERQ at Flotta	Inhalation of smoke and gas
78.	KIRBY, JOHN BRIAN (51) 3 Wheatlands Drive Easington Saltburn Cleveland	Production Operator	Occidental Petroleum (Caledonia) Ltd 1 Claymore Drive Bridge of Don Aberdeen	Recovered 26.10.88 in D Deck of ERQ at Flotta	Inhalation of smoke and gas
79.	KNOX, STUART GORDON CHARLES (37) 119 Morrison Street Edinburgh	Roustabout	John Wood Group plc John Wood House Greenwell Road East Tullos Aberdeen	Recovered 26.10.88 in D Deck of ERQ at Flotta	Inhalation of smoke and gas
80.	LAING, ALEXANDER RODGER (38) 43 Portree Avenue Broughty Ferry Dundee	Steward	Kelvin Catering Camps Ltd 5 Queen's Terrace Aberdeen	Missing	N/A
81.	LARGUE, TERENCE MICHAEL (34) 24 Beale Close Ingleby Barwick Stockton Cleveland	Scaffolder	Aberdeen Scaffolding Company Ltd Hareness Road Aberdeen	Recovered 25.10.88 in D Deck of ERQ at Flotta	Inhalation of smoke and gas
82.	LAWRIE, GRAHAM (39) 146 Faulds Gate Aberdeen	Roustabout	Bawden International Ltd Wellheads Road Dyce Aberdeen	Recovered 26.10.88 in D Deck of ERQ at Flotta	Inhalation of smoke and gas
83.	LEGGAT, FINDLAY WALLACE (37) (formerly CAMPBELL) 15/8 Bernard Terrace Bridgeton Glasgow	Scaffolder	Aberdeen Scaffolding Company Ltd Hareness Road Aberdeen	Missing: LKW – in White House, possibly after 22.45	N/A

	Deceased	Occupation	Employer	Recovery/LKW	Principal Cause of Death
84.	LITHGOW, BRIAN (34) 25 Falcon Avenue Edinburgh	Photographic Technician	Stena Offshore Stena House Westhill Industrial Estate Westhill Aberdeen	Recovered 2.6.89 GMR 340152E, 6484085N	Not ascertained
85.	LITTLEJOHN, ROBERT RODGER (29) 9 Otterson Grove Dalgety Bay Fife	Pipe Fitter	John Wood Group plc John Wood House Greenwell Road East Tullos Aberdeen	Recovered 24.10.88 in D Deck of ERQ at Flotta	Inhalation of smoke and gas
86.	LONGSTAFFE, MARTIN GEORGE (22) 903 Ecclesall Road Sheffield	Logger	Exploration Logging North Sea Ltd Walton Road Dyce Aberdeen	Recovered 6.9.88 in Galley of ERQ	Inhalation of smoke and gas
87.	MAHONEY, WILLIAM RAYMOND (60) 255 Girdleness Road Aberdeen	Steward	Kelvin Catering Camps Ltd 5 Queen's Terrace Aberdeen	Recovered 26.10.88 in D Deck of ERQ at Flotta	Inhalation of smoke and gas
88.	MARTIN, JOHN MORRISON (33) 56 Firhill Alness	Rigger	John Wood Group plc John Wood House Greenwell Road East Tullos Aberdeen	Recovered 25.10.88 in D Deck of ERQ at Flotta	Inhalation of smoke and gas
89.	McBOYLE, SIDNEY IAN (36) 24 Stensall Road Huntingdon York	Motorman*	Bawden International Ltd Wellheads Road Dyce Aberdeen	Recovered 26.10.88 in D Deck of ERQ at Flotta	Inhalation of smoke and gas
90.	McCALL, ROBERT BORLAND (39) 56 Mauchline Road Hurlford Ayrshire	Chief Electrician	Bawden International Ltd Wellheads Road Dyce Aberdeen	Recovered 24.10.88 in D Deck of ERQ at Flotta	Inhalation of smoke and gas
91.	McCULLOCH, JAMES (51) 4 Gregness Gardens Torry Aberdeen	H.V.A.C. Technician	John Wood Group plc John Wood House Greenwell Road East Tullos Aberdeen	Recovered 6.9.88 in Galley of ERQ	Not ascertained

	Deceased	Occupation	Employer	Recovery/LKW	Principal Cause of Death
92.	McDONALD, ALISTAIR JAMES (33) 16 First Avenue Stepps Glasgow	Mechanical Technician	Occidental Petroleum (Caledonia) Ltd 1 Claymore Drive Bridge of Don Aberdeen	Recovered 26.10.88 in D Deck of ERQ at Flotta	Inhalation of smoke and gas
93.	McELWEE, ALEXANDER (45) 29 Westwood Quad Clydebank	Plater	John Wood Group plc John Wood House Greenwell Road East Tullos Aberdeen	Missing	N/A
94.	McEWAN, THOMAS O'NEIL (38) 32 Maple Court Abronhill Cumbernauld	Electrical Chargehand	John Wood Group plc John Wood House Greenwell Road East Tullos Aberdeen	Recovered 24.10.88 in D Deck of ERQ at Flotta	Inhalation of smoke and gas
95.	McGREGOR, WILLIAM GEORGE (48) 11 Johnston Gardens West Peterculter Aberdeen	Leading Steward	Kelvin Catering Camps Ltd 5 Queen's Terrace Aberdeen	Recovered 7.7.88 on surface of sea near Platform	Chest injury
96.	McGURK, FREDERICK THOMAS SUMMERS (51) 131 Fintry Drive Dundee	Rigger	John Wood Group plc John Wood House Greenwell Road East Tullos Aberdeen	Missing	N/A
97.	McINTOSH, WILLIAM HUGH (24) 80 Highfield Walk Turriff	Floorman	Bawden International Ltd Wellheads Road Dyce Aberdeen	Recovered 24.10.88 in D Deck of ERQ at Flotta	Inhalation of smoke and gas
98.	McKAY, GORDON (33) 231 West Road Fraserburgh	Valve Technician	John Wood Group plc John Wood House Greenwell Road East Tullos Aberdeen	Recovered 14.7.88 80m, 044 degrees from Leg A5	Inhalation of smoke and gas
99.	McLAUGHLIN, CHARLES EDWARD (46) 1 Unity Place Woodside Glasgow	Electrician	John Wood Group plc John Wood House Greenwell Road East Tullos Aberdeen	Recovered 7.7.88 on surface of sea near Platform: LKW – going towards SW corner of Platform shortly after 22.40	Drowning

Deceased	Occupation	Employer	Recovery/LKW	Principal Cause of Death
100. McLEOD, NEIL STUART ROSS (41) Doleygate Farm Gnosall Stafford	Quality Assurance Inspector	Leuven Services (Aberdeen) Ltd 3 Albert Street Aberdeen	Recovered 25.10.88 in D Deck of ERQ at Flotta	Inhalation of smoke and gas
101. McPAKE, FRANCIS (49) 54D Woodside Road Raploch Stirling	Steel Erector/Rigger	John Wood Group plc John Wood House Greenwell Road East Tullos Aberdeen	Recovered 24.10.88 in D Deck of ERQ at Flotta	Inhalation of smoke and gas
102. McWHINNIE, DAVID ALLISON (36) 3 Carey Road Saltcoats Ayrshire	Production Operator	M B Services Ltd Unit 4E Wellheads Industrial Estate Dyce Aberdeen	Recovered 24.10.88 in D Deck of ERQ at Flotta	Inhalation of smoke and gas
103. McWILLIAMS, DUGALD McLEAN (31) 29 Mearns Street Greenock	Welder	John Wood Group plc John Wood House Greenwell Road East Tullos Aberdeen	Recovered 26.10.88 in D Deck of ERQ at Flotta	Inhalation of smoke and gas
104. MEARNS, CARL (20) 173 Marshall Wallis Road South Shields	Rigger	Stena Offshore Stena House Westhill Industrial Estate Westhill Aberdeen	Recovered 23.8.88 GMR 340104E, 6484116N	Inhalation of smoke and gas
105. MILLAR, DEREK KLEMENT MICHAEL (32) 8 Salters Terrace Dalkeith Midlothian	Supervisor★	Macnamee Services Ltd Burnside Works Kennerty Mills Road Peterculter Aberdeen	Recovered 22.7.88 104m, 091 degrees from Leg A1: LKW – at Reception after 22.00	Inhalation of smoke and gas
106. MILLER, ALAN DAVID (31) 6 Reid Crescent Kirkwall Orkney	Industrial Chemist	Caleb Brett International Wellheads Crescent Dyce Aberdeen	Recovered 25.10.88 in D Deck of ERQ at Flotta	Inhalation of smoke and gas
107. MILLER, FRANK (33) 6 Train Terrace Rosyth Fife	Scaffolder	Aberdeen Scaffolding Company Ltd Hareness Road Aberdeen	Recovered 6.9.88 in Galley of ERQ	Inhalation of smoke and gas

Deceased	Occupation	Employer	Recovery/LKW	Principal Cause of Death
108. MOLLOY, JOHN HECTOR (32) 67 Marlborough Park South Belfast	Engineer	NL Sperry Sun/MWD/Baroid Logging Systems Wellheads Road Dyce Aberdeen	Missing	N/A
109. MORRIS, LESLIE JAMES (38) Pine Lodge Mill of Lumphart Oldmeldrum	Platform Supt.#	Bawden International Ltd Wellheads Road Dyce Aberdeen	Recovered 7.7.88 on surface of sea near Platform: LKW – going towards SW corner of Platform shortly after 22.40	Drowning
110. MUNRO, BRUCE ALEXANDER FERGUSON (29) 96 Auchmill Road Bucksburn Aberdeen	Floorman	Bawden International Ltd Wellheads Road Dyce Aberdeen	Recovered 24.10.88 in D Deck of ERQ at Flotta	Inhalation of smoke and gas
111. MURRAY, GEORGE FAGAN (37) 8 Ravensby Road Carnoustie	Steward	Kelvin Catering Camps Ltd 5 Queen's Terrace Aberdeen	Recovered 25.10.88 in D Deck of ERQ at Flotta	Inhalation of smoke and gas
112. NIVEN, JAMES COWIE (27) 34 Chisholm Place Grangemouth	Roustabout	John Wood Group plc John Wood House Greenwell Road East Tullos Aberdeen	Recovered 24.10.88 in D Deck of ERQ at Flotta	Inhalation of smoke and gas
113. NOBLE, GRAHAM SIM (37) 25 Broomhill Fraserburgh	Materials Man	Occidental Petroleum (Caledonia) Ltd 1 Claymore Drive Bridge of Don Aberdeen	Recovered 3.11.88 in C Deck of ERQ at Flotta	Inhalation of smoke and gas
114. O'SHEA, MICHAEL (30) 1 Millview Terrace Neilston Glasgow	Electrician	John Wood Group plc John Wood House Greenwell Road East Tullos Aberdeen	Recovered 17.7.88 146m, 67 degrees from Leg A5	Inhalation of smoke and gas
115. PEARSTON, ROBERT RENNIE (25) 63 Colthill Circle Milltimber Aberdeen	Mechanic	Macnamee Services Ltd Burnside Works Kennerty Mills Road Peterculter Aberdeen	Missing: LKW – in Cinema about 22.00	N/A

Deceased	Occupation	Employer	Recovery/LKW	Principal Cause of Death
116. PIPER, IAN (38) 26 Hillswick Road Sheddocksley Aberdeen	Motorman	Bawden International Ltd Wellheads Road Dyce Aberdeen	Recovered 26.10.88 in Reception of D Deck of ERQ at Flotta	Inhalation of smoke and gas
117. POCHRYBNIAK, WASYL (37) 2 Westbourne Park Urmston Manchester	Lead Roustabout*	Bawden International Ltd Wellheads Road Dyce Aberdeen	Recovered 26.10.88 in D Deck of ERQ at Flotta	Inhalation of smoke and gas
118. PRICE, RAYMOND LESLIE (59) 57 Stanhope Drive Bromborough Wirrall	Production Operator	Occidental Petroleum (Caledonia) Ltd 1 Claymore Drive Bridge of Don Aberdeen	Missing: LKW – in Electricians' Workshop and possibly White House between 22.00 and 22.20	N/A
119. PYMAN, NEIL (32) 3 Strathburn Gardens Inverurie	Engineer	Eastman Christensen Ltd Murcar Commercial Park Denmore Road Bridge of Don Aberdeen	Recovered 24.10.88 in D Deck of ERQ at Flotta	Inhalation of smoke and gas
120. QUINN, TERENCE STEPHEN (28) 2 Roseville Road Hayes Middlesex	Service Engineer	Orbit Valve Co Europe Orbit House Swallowfield Way Hayes Middlesex	Recovered 25.10.88 in D Deck of ERQ at Flotta	Inhalation of smoke and gas
121. RAEBURN, WILLIAM WALLACE (38) 2 Papes Cottages Roseburn Edinburgh	Maintenance Controller	Occidental Petroleum (Caledonia) Ltd 1 Claymore Drive Bridge of Don Aberdeen	Recovered 31.10.88 42m, 300 degrees from Leg A5	Inhalation of smoke and gas
122. REID, DONALD (44) 4 Greenbrae Gardens South Bridge of Don Aberdeen	Chargehand Engineer	John Wood Group plc John Wood House Greenwell Road East Tullos Aberdeen	Recovered 7.7.88 on surface of sea near Platform	Inhalation of smoke and gas
123. REID, ROBERT WELSH (27) 37 Grant Road Arbroath	Roustabout	Bawden International Ltd Wellheads Road Dyce Aberdeen	Recovered 25.10.88 in D Deck of ERQ at Flotta	Inhalation of smoke and gas

Deceased	Occupation	Employer	Recovery/LKW	Principal Cause of Death
124. RENNIE, GORDON MacALONAN (52) 162 Avontoun Park Linlithgow West Lothian	Process Operator*	M B Services Ltd Unit 4E Wellheads Industrial Estate Dyce Aberdeen	Missing	N/A
125. RICHARD, ROBERT MILLER (45) 16 Polmont Park Polmont Falkirk	Production Operator*	M B Services Ltd Unit 4E Wellheads Industrial Estate Dyce Aberdeen	Missing: LKW – leaving 68 ft level for C Module shortly before 22.00	N/A
126. RIDDOCH, ALAN (44) 22 Morven Crescent Findochty Buckie	Steward	Kelvin Catering Camps Ltd 5 Queen's Terrace Aberdeen	Recovered 6.9.88 in Galley of ERQ	Inhalation of smoke and gas
127. ROBERTS, ADRIAN PETER (28) 2 Coed-y-Llwyn Gellilydan Gwynedd North Wales	Roughneck	Bawden International Ltd Wellheads Road Dyce Aberdeen	Recovered 25.7.88 31m, 245 degrees from Leg A5: LKW – at door leading from Reception at 22.05	Inhalation of smoke and gas
128. ROBERTSON, ALEXANDER JAMES (50) 14 Ritchie Place Crieff	Lead Production Technician#	Occidental Petroleum (Caledonia) Ltd 1 Claymore Drive Bridge of Don Aberdeen	Recovered 25.10.88 in D Deck of ERQ at Flotta	Inhalation of smoke and gas
129. ROBERTSON, DONALD NICHOLSON (54) 31 Sinclair Drive Largs Ayrshire	Mechanical Technician	Occidental Petroleum (Caledonia) Ltd 1 Claymore Drive Bridge of Don Aberdeen	Recovered 25.10.88 in D Deck of ERQ at Flotta	Inhalation of smoke and gas
130. ROSS, GARY (29) 27 Stafford Street Aberdeen	Roustabout*	Bawden International Ltd Wellheads Road Dyce Aberdeen	Recovered 2.11.88 GMR 340107E, 6484109N: LKW – in Gymnasium after 22.00	Inhalation of smoke and gas
131. RYAN, MICHAEL HECTOR (23) 819 Great Northern Road Aberdeen	Roustabout	Bawden International Ltd Wellheads Road Dyce Aberdeen	Recovered 7.7.88 on surface of sea near Platform	Chest injury

Deceased	Occupation	Employer	Recovery/LKW	Principal Cause of Death
132. SANGSTER, STANLEY (56) 52 Cummings Park Drive Aberdeen	Foreman Scaffolder	Aberdeen Scaffolding Co Ltd Hareness Road Aberdeen	Recovered 24.10.88 in D Deck of ERQ at Flotta	Inhalation of smoke and gas
133. SAVAGE, JAMES JOHN DEARN (41) 8A Millfield Drive Polmont Falkirk	Electrical Technician	Occidental Petroleum (Caledonia) Ltd 1 Claymore Drive Bridge of Don Aberdeen	Recovered 2.11.88 GMR 340107E, 6484109N	Inhalation of smoke and gas
134. SCORGIE, MICHAEL HUGH BRODIE (28) 43 Turnberry Drive Kirkcaldy Fife	Lead Floorman*	Bawden International Ltd Wellheads Road Dyce Aberdeen	Recovered 19.7.88 111m, 086 degrees from Leg A5	Inhalation of smoke and gas
135. SCORGIE, WILLIAM ALEXANDER (46) 11 Brimmond Way Westhill Aberdeen	Pipe Fitter	John Wood Group plc John Wood House Greenwell Road East Tullos Aberdeen	Recovered 24.10.88 in D Deck of ERQ at Flotta	Inhalation of smoke and gas
136. SCOTT, JOHN FRANCIS (26) 1 Crew Loan Edinburgh	Scaffolder	Aberdeen Scaffolding Co Ltd Hareness Road Aberdeen	Missing: LKW – leaving Cinema shortly after 22.02	N/A
137. SEATON, COLIN DENIS (51) Lee Croft Kilburn Thirsk North Yorkshire	Offshore Installation Manager#	Occidental Petroleum (Caledonia) Ltd 1 Claymore Drive Bridge of Don Aberdeen	Recovered 25.10.88 in D Deck of ERQ at Flotta	Inhalation of smoke and gas
138. SELBIE, ROBERT HENDRY (32) 1 Newburgh Circle Bridge of Don Aberdeen	Turbo Drill Engineer	Neyfor (UK) Ltd Hareness Circle Aberdeen	Recovered 23.10.88 in D Deck of ERQ at Flotta	Inhalation of smoke and gas
139. SERINK, MICHAEL JEFFREY (26) 7481 Gatineau Place Vancouver British Columbia Canada	Logger	Exploration Logging North Sea Ltd Walton Road Dyce Aberdeen	Recovered 31.10.88 in A Deck of ERQ at Flotta	Inhalation of smoke and gas

Deceased	Occupation	Employer	Recovery/LKW	Principal Cause of Death
140. SHORT, MICHAEL BERNARD (41) 23 Alfred Road Buckhurst Hill Essex	Foreman Rigger	Stena Offshore Stena House Westhill Industrial Estate Westhill Aberdeen	Recovered 11.7.88 GMR 340066E, 6484188N	Inhalation of smoke and gas
141. SKINNER, RICHARD VALENTINE (41) 27 Baillie Norrie Crescent Montrose	Assistant Driller	Bawden International Ltd Wellheads Road Dyce Aberdeen	Recovered 14.7.88 GMR 340205E, 6484311N: LKW – on top of Accom. W shortly before 22.20	Drowning
142. SMITH, WILLIAM HAMILTON (43) "Crammond" 10 Nobel View Reddingmuirhead Falkirk	Maintenance Lead Hand	Occidental Petroleum (Caledonia) Ltd 1 Claymore Drive Bridge of Don Aberdeen	Recovered 23.10.88 in D Deck of ERQ at Flotta	Inhalation of smoke and gas
143. SPEIRS, JAMES (42) 31 MacFarlane Place Uphall West Lothian	Mechanical Technician	John Wood Group plc John Wood House Greenwell Road Bridge of Don Aberdeen	Missing	N/A
144. STEPHENSON, KENNETH STUART (37) 26 Claudius Court South Shields	Rigger	Stena Offshore Stena House Westhill Industrial Estate Westhill Aberdeen	Recovered 12.7.88 150m 260 degrees from Leg B5	Inhalation of smoke and gas
145. STIRLING, THOMAS CUNNINGHAM BOSWELL (27) 6 Burnbank Gardens Maryhill Glasgow	Cleaner	Macnamee Services Ltd Burnside Works Kennerty Mills Road Peterculter Aberdeen	Recovered 18.7.88 109m 072 degrees from Leg A5: LKW – in Galley after 22.00	Inhalation of smoke and gas
146. STOREY, MALCOLM JOHN (38) 22 Coul Park Alness Ross-shire	Seaman	Haven Shipping Co Queen's Road Great Yarmouth	Recovered 7.7.88 on surface of sea near Platform: LKW – in FRC of Sandhaven on W side of Platform about 22.50	Drowning
147. STOTT, JAMES CAMPBELL (40) 3 Newburgh Circle Bridge of Don Aberdeen	Plumber	John Wood Group plc John Wood House Greenwell Road East Tullos Aberdeen	Recovered 6.9.88 in Galley of ERQ	Not Ascertained

Deceased	Occupation	Employer	Recovery/LKW	Principal Cause of Death
148. STWERKA, JURGEN TILO (36) 6843 Biblis 1 Beethoven Strasse 5 Hessen West Germany	Research Chemist	Grace G.M.B.H. Postfach 1445 In Der Hollerhecke D-6520 Worms West Germany	Recovered 26.10.88 in D Deck of ERQ at Flotta	Inhalation of smoke and gas
149. SUTHERLAND, STUART DOUGLAS (21) 9 Marchbank Road Bieldside Aberdeen	Student/Cleaner*	Macnamee Services Ltd Burnside Works Kennerty Mills Road Peterculter Aberdeen	Missing: LKW – in White House after 22.20	N/A
150. SUTTON, TERRENCE JOHN (28) 18 Clinton Drive Sandhaven Fraserburgh	Mechanical Fitter	Score (UK) Ltd Glenugie Engineering Works Invernettie Peterhead	Missing: LKW — in White House after 22.20	N/A
151. TAYLOR, ALEXANDER RONALD (57) 29 Gairnshiel Place Aberdeen	Roustabout	John Wood Group plc John Wood House Greenwell Road East Tullos Aberdeen	Recovered 24.10.88 in D Deck of ERQ at Flotta	Inhalation of smoke and gas
152. THOMPSON, ALISTAIR ADAM (45) 10 Whitestripes Way Bridge of Don Aberdeen	Telecom Engineer	British Telecom New Telecom House College Street Aberdeen	Recovered 29.7.88 152m, 310 degrees from Leg A5	Inhalation of smoke and gas
153. VERNON, ROBERT ARGO (51) 3 The Orchard Brightons Falkirk	Production Operator*	Occidental Petroleum (Caledonia) Ltd 1 Claymore Drive Bridge of Don Aberdeen	Missing: LKW — on NW Navigation Aid Platform shortly before 22.20	N/A
154. WAKEFIELD, JOHN EDWARD (35) 38 Chadwell Springs Waltham Grimsby Humberside	Instrument Technician	Occidental Petroleum (Caledonia) Ltd 1 Claymore Drive Bridge of Don Aberdeen	Recovered 17.10.88 GMR 340106E, 6484112N adjacent to Leg B5: LKW – leaving Instrument Workshop shortly before 22.00	Drowning
155. WALKER, MICHAEL ANDREW (24) 12 Dunnottar Road Castlepark Ellon	Technician	Wimpol Ltd Offshore Navigation Surveys Unit 6 5 Wellheads Industrial Centre Dyce Aberdeen	Recovered 25.10.88 in D Deck of ERQ at Flotta	Inhalation of smoke and gas

Deceased	Occupation	Employer	Recovery/LKW	Principal Cause of Death
156. WARD, BRYAN THOMAS (49) 168 Clarendon Street Hull	Rigger	Stena Offshore Stena House Westhill Industrial Estate Westhill Aberdeen	Recovered 7.7.88 on surface of sea near Platform	Inhalation of smoke and gas
157. WATKIN, GARETH HOPSON (42) 22 Bain Road Mintlaw Aberdeenshire	Offshore Medical Attendant#	Occidental Petroleum (Caledonia) Ltd 1 Claymore Drive Bridge of Don Aberdeen	Recovered 14.7.88 50m, 245 degrees from Leg B5: LKW – in Accommodation, going to Medics' Room with injured man after 22.00	Inhalation of smoke and gas
158. WATSON, FRANCIS JOHN (38) 67 Highfield Road Longlands Middlesbrough	Head Chef	Kelvin Catering Camps Ltd 5 Queen's Terrace Aberdeen	Missing	N/A
159. WHIBLEY, ALEXANDER (28) 1 Crookmore Cottages Montgarrie Alford	Roustabout*	Bawden International Ltd Wellheads Road Dyce Aberdeen	Recovered 25.10.88 in D Deck of ERQ at Flotta	Inhalation of smoke and gas
160. WHITE, KEVAN DENNIS (42) 17/18 Everingham Orton Brimbles Peterborough	Maintenance Supervisor#	Occidental Petroleum (Caledonia) Ltd 1 Claymore Drive Bridge of Don Aberdeen	Recovered 20.7.88 110m, 246 degrees from Leg A5	Chest Injury
161. WHITELEY, ROBERT (39) 17 Queen Margaret Drive Glenrothes Fife	Roustabout	John Wood Group plc John Wood House Greenwell Road East Tullos Aberdeen	Recovered 19.7.88 89m, 078 degrees from Leg A5: LKW – at Reception at 22.05	Inhalation of smoke and gas
162. WHYTE, GRAHAM GILL (42) 17 Carron Place Aberdeen	Aerial Rigger	British Telecom New Telecom House College Street Aberdeen	Recovered 23.7.88 65m, 79 degrees from Leg A5	Death in fire
163. WHYTE, JAMES GILBERT (53) 14 Cramond Park Edinburgh	Aerial Rigger	British Telecom New Telecom House College Street Aberdeen	Recovered 20.7.88 174m, 050 degrees from Leg A5	Inhalation of smoke and gas

483

Deceased	Occupation	Employer	Recovery/LKW	Principal Cause of Death
164. WICKS, ALAN (40) 72 Chapman Drive Carnoustie	Safety Supervisor#	Occidental Petroleum (Caledonia) Ltd 1 Claymore Drive Bridge of Don Aberdeen	Recovered 20.7.88 80m, 278 degrees from Leg A5: LKW – near W door on C Deck shortly before 22.20	Death in fire
165. WILLIAMSON, PAUL CHARLES FERGUSON (24) 8 Fairview Wynd Danestone Aberdeen	Floorman*	Bawden International Ltd Wellheads Road Dyce Aberdeen	Recovered 23.10.88 in D Deck of ERQ at Flotta	Inhalation of smoke and gas
166. WISER, DAVID (65) 140 Drayton Park Highbury London	Survey Technician	UDI Group Ltd Offshore Instrumentation Denmore Road Bridge of Don Aberdeen	Recovered 25.10.88 in D Deck of ERQ at Flotta	Inhalation of smoke
167. WOODCOCK, JOHN RICHARD (29) 9 Claremont Street Aberdeen	Technical Clerk	John Wood Group plc John Wood House Greenwell Road East Tullos Aberdeen	Missing: LKW — in Tea Hut above pipe deck after 22.20	N/A

NOTES: No 10 (Brian Batchelor) and No 146 (Malcolm Storey) were members of the crew of the Fast Rescue Craft of the *Sandhaven*.

The symbol * denotes that the deceased was on night shift. The symbol # indicates that he was on 24 hours call.

The column "Recovery/LKW" sets out information as to the recovery of the body of the deceased where this was achieved. In the case of those missing and those recovered from elsewhere than the wreckage of the accommodation it sets out the last known whereabouts (LKW) of the deceased in the period from about 22.00 on 6 July in the light of the evidence available to the Inquiry.

Appendix I
List of Abbreviations

Acronyms

AAE	Additional Accommodation East
AAW	Additional Accommodation West
AC	Alternating current
ACMH	Advisory Committee on Major Hazards
AFE	Authorisation for expenditure
AIT	Autoignition temperature
ALARP	As low as reasonably practicable
APAU	Accident Prevention Advisory Unit (of HSE)
ASTM	American Society for Testing and Materials
BA	Breathing apparatus
BOP	Blowout preventer
BS	British Standard
BST	British Standard Time
CAA	Civil Aviation Authority
CBI	Confederation of British Industry
CIMAH	Control of Industrial Major Accident Hazards (Regulations)
COSHH	Control of Substances Hazardous to Health (Regulations)
CSE	Concept safety evaluation
CSS	Co-ordinator surface search
DC	Direct current
DEn	Department of Energy
DHSV	Down hole safety valve
DoT	Department of Transport
DOTI	Department of Trade and Industry
DPCV	Differential pressure control valve
DSF	Deck support frame
EADU	Exploration, Appraisal and Development Unit (of DEn)
ECV	Emergency control valve
EEC	Emergency Evacuation Controller
EPS	Emergency power supply
ERQ	East Replacement Quarters
ESD	Emergency shutdown
ESV	Emergency shutdown valve
FCV	Flow control valve
F&G	Fire and gas (panel, system)
FILO	Flight information and logistics officer
FRC	Fast rescue craft
FSA	Formal safety assessment
GA	General alarm
GCM	Gas Conservation Module
GMT	Greenwich Mean Time

GOV	Gas operated valve
GTC	Gas to Claymore (valve)
HAZOP	Hazard and operability (study)
HF	High frequency
HFT	Hydrate formation temperature
HMV	Hydraulic master valve
HP	High pressure
HSC	Health and Safety Commission
HSE	Health and Safety Executive
HSWA	Health and Safety at Work etc, Act 1974
HVAC	Heating, ventilation and air conditioning
IMO	International Maritime Organisation
IP	Institute of Petroleum
IR	Infra-red
IUOOC	Inter-Union Offshore Oil Committee
JB	John Brown (generators)
JCP	Joint control panel
JT	Joule Thomson (effect, flash drum, valve)
LCV	Level control valve
LEL	Lower explosive limit
LIC	Level indicator controller
LP	Low pressure
LPG	Liquefied petroleum gas
LQW	Living Quarters West
LSA	Low specific activity
MCP	Manifold compression platform
MERSAR	Merchant Ship Search and Rescue (Manual)
MHAU	Major Hazards Assessment Unit (of HSE)
MOB	Man overboard
MOL	Main oil line
MOV	Motor operated valve
MRCC	Maritime Rescue Co-ordination Centre
MWA	Mineral Workings (Offshore Installations) Act 1971
MWP	Maximum working pressure
NCS	Norwegian continental shelf
NEL	National Engineering Laboratory
NPD	Norwegian Petroleum Directorate
NRV	Non-return valve (also called check valve)
OESD	Overall emergency shutdown
OIAC	Offshore Industry Advisory Committee
OIM	Offshore Installation Manager
OPG	Offshore Projects Group
OSC	On-Scene Commander
PA	Personal address
PCV	Pressure control valve
PED	Petroleum Engineering Division (of DEn)
PESD	Platform emergency shutdown

PM	Preventive maintenance
POB	Persons on board
PPA	Petroleum (Production) Act 1934
PSPA	Petroleum and Submarine Pipe-lines Act 1975
PSV	Pressure safety valve
PTW	Permit to work
QA	Quality assurance
QMS	Quality management system
QRA	Quantitative risk assessment
RCC	Rescue co-ordination centre
RGIT	Robert Gordon Institute of Technology
RIV	Rapid intervention vessel
RNLI	Royal National Lifeboat Institution
ROV	Remotely operated vehicle
RTJ	Ring type joint
SBV	Standby vessel
SMS	Safety management system
SOLAS	Safety of Life at Sea
SPEEM	Submersible Pump Electrical Equipment Module
SRV	Safety relief valve
SSIV	Subsea isolation valve
TEMPSC	Totally enclosed motor propelled survival craft
TRA	Total risk analysis
TSR	Temporary safe refuge
TUC	Trades Union Congress
UHF	Ultra high frequency
UKCS	United Kingdom continental shelf
UKOOA	United Kingdom Offshore Operators Association Ltd
UPS	Uninterrupted power supply
UV	Ultra-violet
VDU	Visual display unit
VHF	Very high frequency

Units

BPD	barrels per day
MCF	thousands of cubic feet
MMSCF	millions of standard cubic feet
MMSCFD	millions of standard cubic feet per day
SCF	standard cubic feet
bara	bars absolute
barg	bars gauge
bbl/d	barrels per day
cfm	cubic feet per minute
psia	pounds per square inch absolute
psig	pounds per square inch gauge
scfm	standard cubic feet per minute
v/v	volume/volume
w/w	weight/weight

Statutory Instruments

Construction and Survey Regulations - Offshore Installations (Construction and Survey) Regulations 1974 (SI 1974 No 289)

Emergency Pipe-line Valve Regulations - Offshore Installations (Emergency Pipe-line Valve) Regulations 1989 (SI 1989 No 1029)

Emergency Procedures Regulations - Offshore Installations (Emergency Procedures) Regulations 1976 (SI 1976 No 1542)

Fire-fighting Equipment Regulations - Offshore Installations (Fire-fighting Equipment) Regulations 1978 (SI 1978 No 611)

Included Apparatus or Works Order - Offshore Installations (Included Apparatus or Works) Order 1989 (SI 1989 No 978)

Inspectors and Casualties Regulations - Offshore Installations (Inspectors and Casualties) Regulations 1973 (SI 1973 No 1842)

Life-saving Appliances Regulations - Offshore Installations (Life-saving Appliances) Regulations 1977 (SI 1977 No 486)

Operational Safety, Health and Welfare Regulations - Offshore Installations (Operational Safety, Health and Welfare) Regulations 1976 (SI 1976 No 1019)

Public Inquiries Regulations - Offshore Installations (Public Inquiries) Regulations 1974 (SI 1974 No 338)

Well Control Regulations - Offshore Installations (Well Control) Regulations 1980 (SI 1980 No 1759)

Printed in the United Kingdom for Her Majesty's Stationery Office
Dd 0503434/11/90, C75

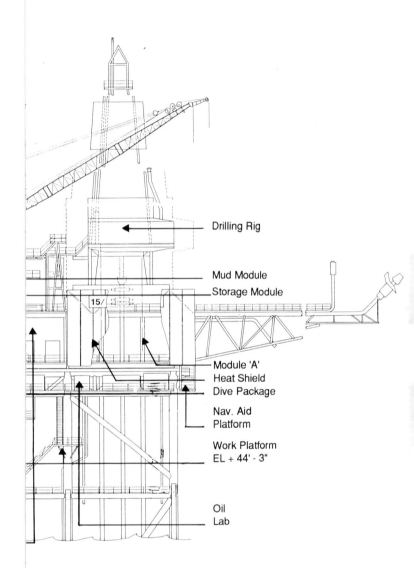

Drilling Rig

Mud Module

Storage Module

15/

Module 'A'
Heat Shield
Dive Package

Nav. Aid
Platform

Work Platform
EL + 44' - 3"

Oil
Lab

)RM WEST ELEVATION

b)

Helideck

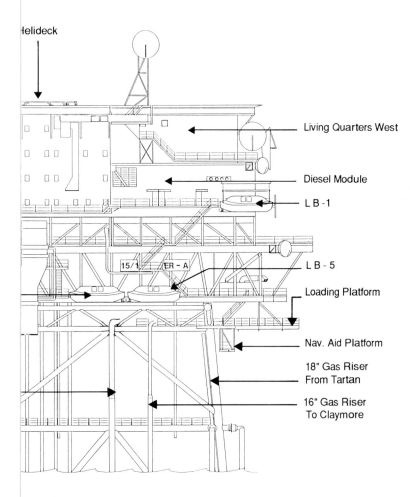

Living Quarters West

Diesel Module

L B - 1

15/1... ER - A

L B - 5

Loading Platform

Nav. Aid Platform

18" Gas Riser
From Tartan

16" Gas Riser
To Claymore

PLATFORM NORTH ELEVATION

(b)

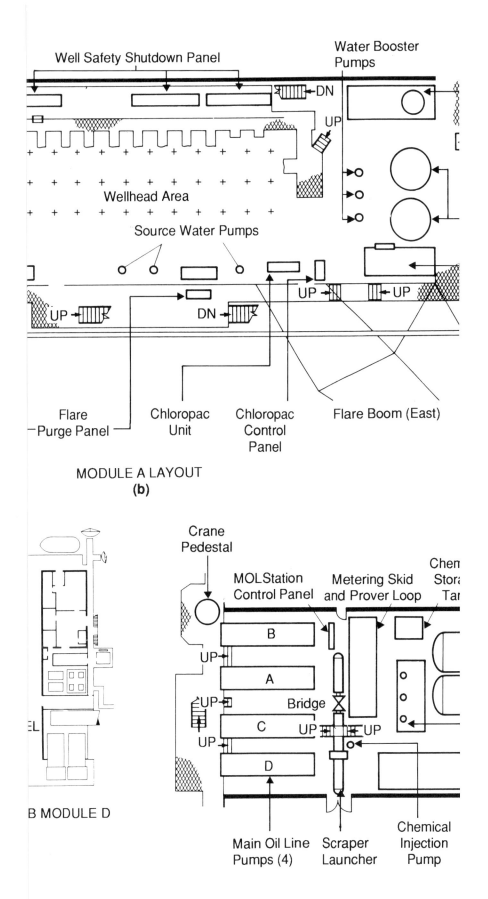

Well Safety Shutdown Panel

Water Booster Pumps

DN

UP

Wellhead Area

Source Water Pumps

UP

DN

UP — UP

UP

Flare
Purge Panel

Chloropac
Unit

Chloropac
Control
Panel

Flare Boom (East)

MODULE A LAYOUT
(b)

Crane
Pedestal

MOLStation
Control Panel

Metering Skid
and Prover Loop

Chem
Stora
Tar

B

UP

A

UP

Bridge

C   UP   UP

UP

D

B MODULE D

Main Oil Line
Pumps (4)

Scraper
Launcher

Chemical
Injection
Pump

MOD

production deck modules — 1: (a) production deck; (b) A Module; a

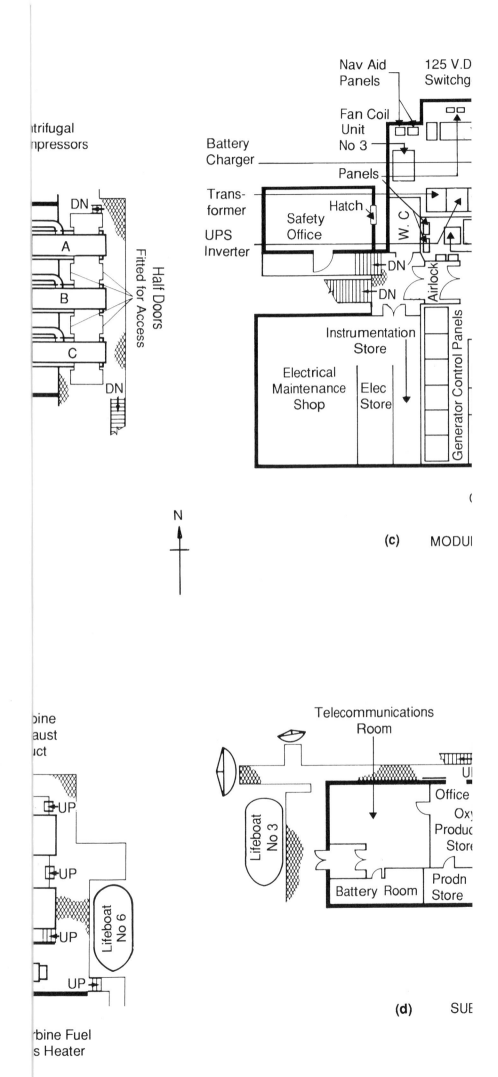

N

(c) MODU

(d) SUB

tion deck and production deck modules — 2: (a) C Module; (b) D Mo
hine level; and (d) Submodule D.

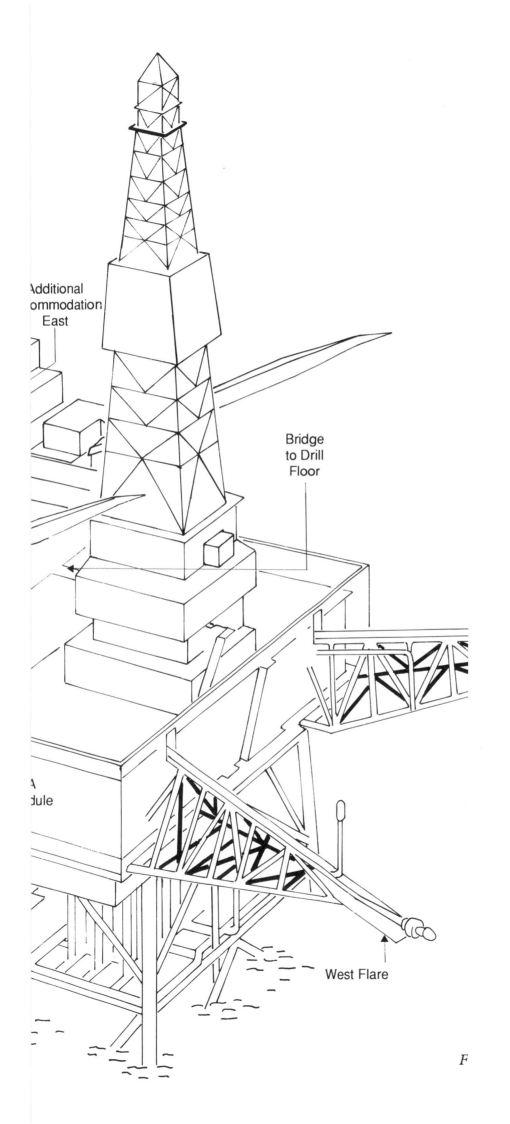

Additional
Accommodation
East

Bridge
to Drill
Floor

A
Module

West Flare

F

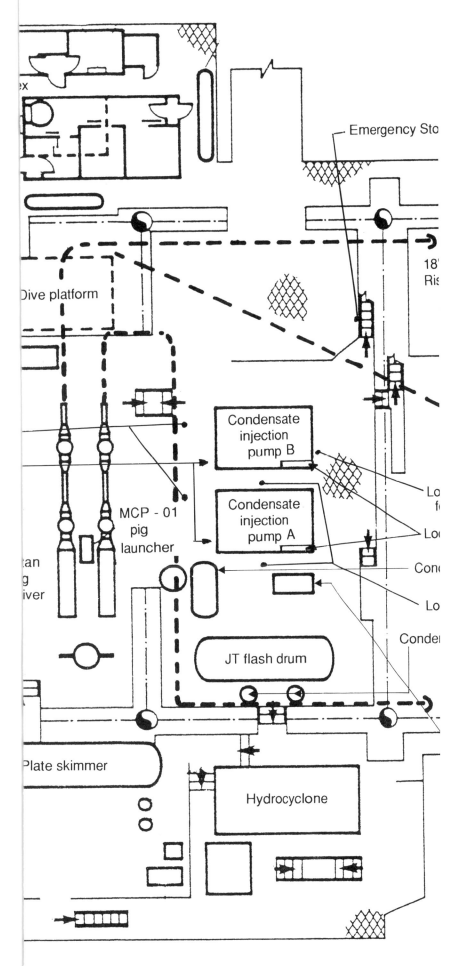

18'
Ris

Dive platform

Condensate
injection
pump B

Condensate
injection
pump A

MCP - 01
pig
launcher

tan
g
iver

JT flash drum

Lo
f
Lo
Con
Lo
Conder

Plate skimmer

Hydrocyclone

y. J.6   The 68ft level, or deck support frame.

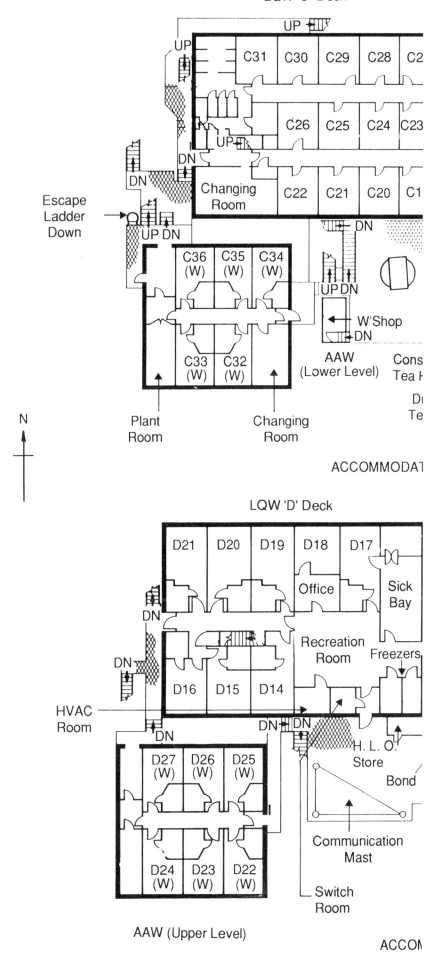

Fig. J.7 Layout of the accommodation modules: (a) level 1 (ERQ A
(ERQ and LQW C deck, AAE upper level and AAW lowe
level).

230 Vapour	240 Vapour	30 Liq	700 Vapour	710 Vapour	720 Vapour	730 Vapour
198	198	64	64.2	105	55.6	198
1735	1735	15	150	675	635	1,735
1.000	1.000	C	1.000	1.000	1.000	1.000
22.6	22.6	45	25.6	25.1	22.6	22.6
6.61	6.61	46	0.721	3.42	3.19	6.61
–	–	C	–	–	–	–
132,720	132,670	22	8,431	45,411	12,598	46
5,877	5,875	499	329.4	1,812	557.9	2
–	–	2	–	–	–	–
53.5	53.5	–	3.0	16.5	5.1	0.0
2.7	2.7	237	0.7	3.5	0.3	0.0
0.19	0.19	C	0.01	0.06	0.02	0.00
24.6	24.6	C	1.3	7.2	2.3	0.0
252.3	252.2	C	12.2	68.3	23.9	0.1
3,989.8	3,988.4	9	198.7	1,112.2	378.7	1.4
916.1	915.8	14	52.9	291.7	87.0	0.3
547.0	546.8	41	41.9	224.5	51.9	0.2
29.5	29.4	7	3.2	16.4	2.8	0.0
85.7	85.7	39	11.0	55.1	8.1	0.0
14.2	14.2	27	3.1	14.0	1.3	0.0
12.5	12.5	39	3.2	14.2	1.2	0.0
2.5	2.5	45	1.3	4.7	0.2	0.0
0.0	0.0	36	0.1	0.1	0.0	0.0
0.0	0.0	C	0.0	0.0	0.0	0.0
0.0	0.0	C	0.0	0.0	0.0	0.0
0.0	0.0	C	0.0	0.0	0.0	0.0
0.0	0.0	C	0.0	0.0	0.0	0.0

T V
HAS

200

190

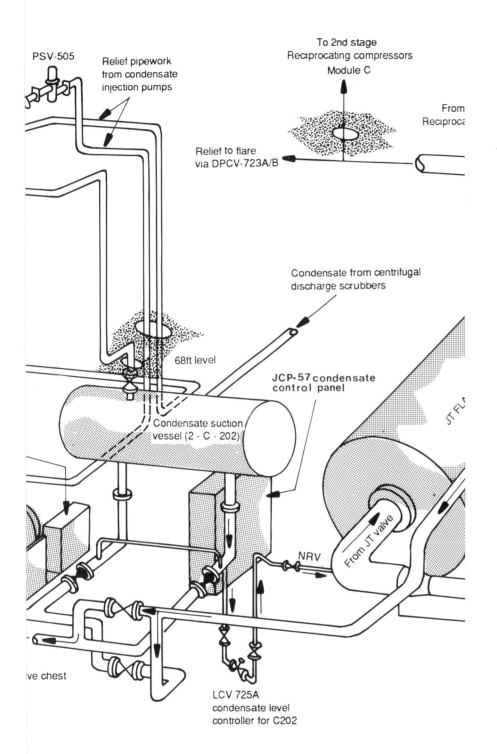

PSV-505

Relief pipework
from condensate
injection pumps

To 2nd stage
Reciprocating compressors
Module C

From
Reciproca

Relief to flare
via DPCV-723A/B

Condensate from centrifugal
discharge scrubbers

68ft level

JCP-57 condensate
control panel

Condensate suction
vessel (2 - C - 202)

JT FL

From JT valve

NRV

ve chest

LCV 725A
condensate level
controller for C202